T0189730

Indoor Location-Based Services

Martin Werner

Indoor Location-Based Services

Prerequisites and Foundations

 Springer

Martin Werner
Ludwig-Maximilians-Universität München
Munich
Germany

ISBN 978-3-319-35358-6   ISBN 978-3-319-10699-1 (eBook)
DOI 10.1007/978-3-319-10699-1
Springer Cham Heidelberg New York Dordrecht London

Printed on acid-free paper

Springer is part of Springer Science+Business Media (www.springer.com)

To my family.

Preface

The rise of powerful smartphones equipped with many sensors has led to a huge amount of location-based services for the outside world. These applications provide orientation, navigation, and information to users in unknown locations. However, these services have long been limited to the outside of buildings, either because GPS is unavailable inside buildings or because alternative positioning based on cell towers or Wi-Fi access points is too inaccurate. The high information value of these applications, however, has triggered a lot of research towards extending the service experience from the outside into the inside of buildings. From a computer science perspective, this has led to a beautiful area of research in which many previously isolated domains converge towards new ideas and results. This book intends to bring together some of these aspects for students and researchers in a way, such that the aesthetics of indoor location-based services becomes evident. However, writing this book was quite challenging: It is sometimes hard to correctly distinguish the relevant and the irrelevant. However, I really enjoyed writing this book, and I hope that it will be helpful in making the domain of indoor location-based services accessible and attractive for others. Due to the chosen approach of covering the basics of many previously isolated domains in one book, this book cannot serve and is not intended to be a reference or state-of-the-art review. Instead, it provides readers with the needed background to enter the research area themselves.

This book would not have been possible without the help from my colleagues, my friends, and my family and without a university, which provides me with the freedom of investing a lot of time into this project. For their outstanding support and their helpful comments and discussions, I want to thank (in alphabetical order) Ulrich Bareth, Michael Beck, Florian Dorfmeister, Michael Dürr, Sebastian Feld, Moritz Kessel, Helge Klimek, Hans-Peter Kriegel, Claudia Linnhoff-Popien, Marco Maier, Philipp Marcus, Chadly Marouane, Valentin Protschky, Peter Ruppel, Lorenz Schauer, Mirco Schönfeld, and Kevin Wiesner.

Munich, Germany   Martin Werner
May 2014

Contents

Chapter 1
Introduction

A journey of a thousand miles begins with a single step.

Laozi

Positioning and navigation as well as their related areas of mapping and guidance have been a central and widely recognized problem for human societies for centuries. Still, a lot of efforts are made towards more precise, reliable, and ubiquitous positioning systems. Nowadays, GPS and other navigation satellite systems are sufficient for navigating a car inside the complete world. The limitations, that GPS does not work inside tunnels and buildings, are not severe to that application. However, the widespread use of smartphones equipped with GPS leads to a lot of smart services using location as ingredient, and soon the aim for providing better service quality inside buildings comes up. Up to now, there is no system comparable to GPS for the inner of buildings. This is due to the fact that global radio-based infrastructures such as GPS cannot be observed with sufficient quality inside buildings. This is due to the fact that radio propagation of high-frequency signals in buildings faces a lot of multipath situations in which the propagation path is not the direct line between the mobile device and a base station or satellite which is a central assumption for GPS. On the other hand, low frequencies can penetrate buildings on the direct line of sight. However, they cannot be sent from satellites and only provide coarse time information and location. Therefore, a lot of local techniques based on a multitude of different technologies have been proposed to reach the same service quality inside buildings, which is offered outside using GPS. Some systems reach incredible accuracy of localization including recent UWB-based systems. However, they are often limited by their complexity and cost. Other systems reach global availability such as indoor positioning based on Wi-Fi measurements or cellular networks. However, their accuracy is often problematic for providing sufficient service quality. Note that the accuracy demands of personal navigation systems inside buildings are often claimed in the order of 1 or 2 m. This would suffice to distinguish doors located next to each other and enable a person to read off door signs located inside the error radius easily. In contrast to that, the expected error of GPS is around 10 m, a much higher value. Hence, even when a system with performance comparable to GPS is made available inside buildings, this does not imply that location-based services become easily possible. This is one of the main motivations for treating indoor location-based services fundamentally

© Springer International Publishing Switzerland 2014
M. Werner, *Indoor Location-Based Services*, DOI 10.1007/978-3-319-10699-1_1

differently and one of the main reasons for writing a book dedicated to the indoor situation. The high amount of uncertainty of location to be expected inside buildings brings a much higher influence of map information, filtering, and reasoning for providing acceptable overall service quality as compared to the outside. It is to be expected that fulfilling the demand for positioning with errors below 1 m is not possible without integrating map information and probabilistic techniques. This is quite different for GPS navigation: for GPS, a closed system with a single standardized interface and with very limited data access provides location estimates with errors in the magnitude of 10 m. A basic GPS system uses only almanac data defining the current location of satellites, and this data is transmitted via a satellite link. Therefore, no terrestrial network access is needed for basic GPS positioning.

This isolation of the positioning system and the location-based service is impossible for indoor location-based services. One reason for that is that the coarse position returned by simple GPS leads to high-quality locations using pretty simple map matching especially for the application domain of vehicle navigation.

To illustrate this, consider the following example of a car navigation system. The positioning system provides location fixes from time to time, while the navigation backend is interested in the current street segment inside a map of streets. Consider the situation of Fig. 1.1.

A car is driving upwards on the middle road labeled C, and GPS locations with large errors are shown.

A simple map matcher based on tracking multiple hypotheses of the current street segment over time is described as follows: from the first measurement, three hypotheses will be generated. The car is driving on A, B, or C, and these are scored according to their distance to the fix and their previous weighting. Hence, A and C are scored higher from this fix compared to B. For newly arriving measurements 2, 3, and 4, these hypotheses are updated, and possibly, new hypotheses are generated. The hypothesis of lowest score is dropped when the number of hypotheses exceeds a preconfigured limit. Altogether, this deals successfully with GPS uncertainty in most cases as the number of possible movements is low enough to track each individual possible movement due to the low degree of the street network and the small number of streets within the uncertainty of GPS fixes.

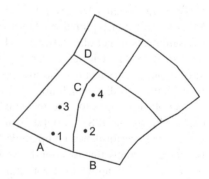

Fig. 1.1 Map matching
example for car navigation

In contrast to that and even when assuming a perfect positioning system without any errors, the indoor area contains a lot of equivalent ways and free space leading to completely different traces of coordinates. Tracking all possible indoor movements quickly becomes infeasible. There is an approach called particle filter, which essentially tries to accomplish this. However, particle filtering applied to some indoor area has to track thousands of particles in order to be able to keep track of most sensible hypotheses. Furthermore, the inherent symmetries of buildings render the removal of hypotheses more difficult and lead to the huge amount of hypothetical locations.

In the last decades, indoor location-based services have often been reduced to solving the indoor positioning problem. While this is one of the most important problems for indoor navigation, it is not the only one, and this makes this book different from many other books dedicated to positioning. This book is intended to also shed light onto the problems waiting in the background when positioning has been solved completely. Moreover, it is to be expected that the problems related to indoor location-based services cannot be solved in isolation as was possible for the outside space, and the book will show examples on how to integrate map information or movement constraints into position determination.

Another perspective on location-based services is given by looking into the organization and communication of devices with location awareness. This includes communication patterns for different location-based service classes. From this perspective, the transmission, coding, storage, and retrieval of location-based information and the model of interaction including push and pull services are discussed. However, this area of research is very similar for location-based services indoors as well as outdoors. Though indoor geolocation is basically more uncertain, optimizations with respect to energy consumption and privacy render outdoor location-based services into similarly uncertain by leaving out available information such as shutting down the GPS to save energy in specific situations. Complicated middleware constructions have been long discussed including the Tracking and Exchange platform which might serve as an example on this area of research.

For the purpose of this book and in order to keep the description concise, we will omit details of service management, deployment, and energy optimization in cases where they do not differ too much from the treatment of these topics for general location-based services. Furthermore, we restrict the discussion of indoor location-based to the most advanced example of navigation. This is based on the observation that a system that is able to provide turn-by-turn guidance to a mobile device including map information, semantic models of the surroundings, and the associated capabilities of a geographic information system (GIS) is sufficient to provide the most relevant other classes of location-based services.

Overall, this book intends to collect results and ideas from very different domains, which are all relevant to providing indoor location-based services in the future.

1.1   Location-Based Services (LBS)

Location-based services are based on the need of people to orient in unknown environments. A location-based service is a computer service, which provides a specific functionality based on the current location of a mobile entity. In some cases, the mobile entity is a mobile device whose location is identified with the location of the person using the device. In other cases, the location of some objects is to be determined and tracked, for example, in fleet management applications or logistics.

There are several definitions of location-based services in literature. In essence, as we will observe in the following examples, location-based services contain three aspects: location, communication, and geoinformation.

Early definitions stem from the telecommunication standardization and usually bring the value of location into the definition. The GSM Association, for example, defines LBS roughly as a service using the location of a target in order to add value to a service. Note that the target is not necessarily a person but is defined to be the entity whose location is to be inferred. This definition includes the aspect of location explicitly. The aspect of communication is included implicitly as the GSM Association is a standardization gremium for mobile communication networks. The geoinformation aspect is given by the requirement of adding value to a service. It is clear that a service that adds value based on positioning needs to derive service information from location.

A more restrictive definition from the research domain reads a bit different as it—unfortunately—restricts location-based services to services in which the service users location is being used:

> In the general case, the location-based services can be defined as services utilizing the ability to dynamically determine and transmit the location of persons within a mobile network by the means of their terminals [6].

This definition also contains the three basic aspects: Firstly, location-based services are services utilizing the ability to dynamically determine the location of persons. This part of the definition is related to the aspect of location. The aspect of communication—though slightly restricted to the communication of location, actually—is included, too. Finally, the geoinformation aspect or the value aspect is included through the word "utilizing."

In order to further illustrate location-based services from a more concrete perspective, consider the following three abstract questions whose answer is utilizing location information:

- *Where am I?*
  The problem of finding a location of the mobile device.
- *What is surrounding me?*
  The problem of giving sense to the inferred location depending on the surroundings.
- *How do I get somewhere?*
  The problem of navigation and guidance in unknown environments.

For the purpose of this book, we will define location-based services as follows: *A location-based service is defined to be a service relying on the following three aspects: the ability to **infer the location** of one or more mobile entities, the ability to **communicate** information, and the ability to **use location data** in order to provide the service.*

This definition is perfectly aligned with the location-based service triangle which can be found in several discussions of location-based services and which is depicted in Fig. 1.2.

Sometimes it is reasonable to fix some roles in location-based services in order to better understand the preconditions and interactions of different entities when designing and developing location-based services. Four such roles are of sufficient universality to further illustrate definitions of location-based services. From this perspective, location-based services are services constructed from the concrete instantiation of systems accomplishing the following tasks:

* *Terminal:* A mobile terminal providing the ability to change location.
* *Location Enabler:* A system enabling the inference of the location of mobile terminals.
* *Service Provider:* A service provider which can generate additional value by utilizing location information
* *Service User:* A user which utilizes the service.

All of these systems can be deployed in different locations and domains. For example, a cellular network can provide a location enabler inside an infrastructure, while a GPS chip can be seen as a location enabler for a smartphone. A terminal can be a smartphone or tabled—the most common case—or an embedded system comprised of nothing more than being mobile and being identified such as a passive RFID tag. A service provider is the entity which is interested in exploiting available location information in order to provide a service, and the service user is the entity using the service.

In practice, a lot of additional roles come up in different scenarios; however, these four roles are always needed to build a location-based service.

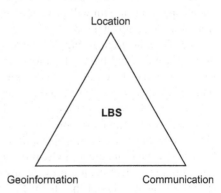

Fig. 1.2 The location-based service triangle

1.2   What Is Special About Indoor LBS

The success of location-based services outside has led to the desire for extending this service experience into the inner of buildings. However, a lot of open problems have to be solved, before a comparable service quality can be achieved inside buildings and position determination (e.g., an equivalent to GPS outside) is only one of them. For the indoor area, the complete value chain of location-based services needs to be addressed, and the diversity of stakeholders as compared to the outside world is often challenging. For example, a simple navigation application needs access to sufficient map information about a building. While it is possible to buy high-quality navigation datasets covering the outside world from a vehicle navigation perspective including streets, ferries, and point-of-interest information from map companies, such a company is not yet available for mapping indoor areas. This is linked to the complexities of creating and maintaining indoor maps as compared to outdoors: outdoor maps can be created from satellite imagery as well as from traces collected using GPS. For the indoor area, however, no such accessible and automated source of information is available, today. The only source of information is often given by a floorplan or drawings used during construction which are full of symbolism and difficult to understand for a computer service. Furthermore, indoor location-based services need a position determination service of sufficient accuracy. In many cases, this can only be provided in cooperation with the building owner unless crowdsourcing techniques have evolved to provide and organize map information.

This whole situation leads to a very fragmented landscape of isolated applications. As it is still impossible to globally deploy an indoor location-based service using some standardized data source, it becomes less likely for actual application providers to accept limitations or additional overhead induced by adopting a proposed standard. Therefore, each application does similar things in quite different ways each time providing the best relation between modeling effort and achievable service quality.

Moreover, mobile platforms tend to limit applications access to sensitive data which clearly includes all data which might enable positioning or other forms of context awareness. This data is to be considered private, and hence a modern smartphone operating system will provide tools and mechanisms to prevent applications from using this information.

The block diagram in Fig. 1.3 depicts a minimal set of components needed to provide a nontrivial indoor location-based service. In comparison to the outside world, the interrelation of different entities involved in providing a location-based service is much more complicated in many cases:

One large building block of indoor location-based services is about the map information and semantic representation of buildings. There is not even a suitable global coordinate system in which indoor location-based services can be performed. Rather, any environmental model provides an own coordinate system. As a result, a positioning system cannot provide any location information without referring to an environmental model in which the location can be interpreted. This relation

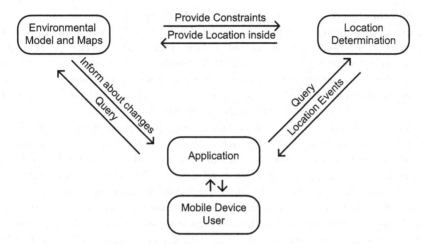

Fig. 1.3 Building blocks of indoor location-based services

between environmental model and positioning becomes critical for positioning systems, which use map-matching techniques to overcome inherent inaccuracies in position determination and for coordinate-free positioning systems. Coordinate-free solutions might provide a room name or some other semantic labels describing the use of the room and possibly a graph describing their topological relationships. This type of location can be inferred much more reliable and is of more use for an application than a set of numbers representing an uncertain three-dimensional spatial location inside an environmental model consisting of a three-dimensional scene composed of solid objects. Moreover, the operation principle of the application is a central question for indoor location-based services. As some positioning systems are based on a dedicated, distributed infrastructure, this infrastructure can take over active parts and, for example, signal the application with a message, when a given situation occurs in the location data or environmental model. This type of service is often called *proactive* service, as the service gets activated only when a situation of interest is detected in the background and independently from the user. More classical services get active on user request. These are called *reactive* services and can provide the integration of environmental models and positioning through the application itself. From a user perspective as well as for energy efficiency inside a mobile device, proactive services are more interesting than reactive services. However, they are also more difficult to realize: they depend on a flexible and powerful interconnection of services from different domains and on mechanisms rating the relevance of location events with respect to the user or a group of users which is a hard problem for itself.

In a classical indoor location-based service setup, the building owner might provide some infrastructure such as Wi-Fi access points and some map information. Then a service on the Internet might provide the mapping of Wi-Fi sightings to locations inside these maps to applications. For a proactive service, however, some online entity has to know and keep track of the current situation and interest of the user. Therefore, the user device must be able to track the current situation and provide this information to the online service or the online service provides its service in a generic way, and the burden of selecting relevant events is put onto the mobile device. Both approaches have their advantages and drawbacks: when the application provides a backend service with virtually unlimited storage and computation with information about the situation of a mobile user, the backend service can assess the relevance of location events with high accuracy and precision and provide a very good service experience. However, a lot of private data has to be transmitted, stored, and retrieved at the online service which is harmful to privacy as well as to the energy consumption as all this information has to be transferred via a mobile link to the backend system. Contrarily, the location service might be quite generic and provide—for example—all interesting events for a complete city in order not to collect all this information used in the first example. Then, however, the service quality is limited by the computational capabilities and the available bandwidth to the mobile device. The trade-off between both approaches can be understood as a trade-off between the number of events versus the number of events relevant to a specific user. When a high fraction of available events is being used by the location-based service, it might become feasible to transmit the information to the mobile device. If, contrarily, most events do not actually affect a single user, then this should be detected on the backed side of the service and not result in useless communication between the online service and the mobile user.

Another aspect why indoor location-based services differ very much from location-based services outside buildings is the complexity of guidance and navigation. While for many applications outside buildings, a graph of nodes connected by edges is used to represent the network of streets, the inner of buildings is composed of free space limited by constraints. In these cases, finding short paths between locations and goals leads to poor results scraping along walls. Hence, an additional step of either the building of a navigation graph in which shortest paths have acceptable quality or a post-optimization heuristic for shortest paths is needed.

To clarify this issue, Fig. 1.4 depicts a true shortest path drawn as a solid line. It touches the first box and scrapes along the lower bound of the second box and is clearly the shortest path. However, the dotted path has much better visualization properties and is equivalent to the shortest path in the sense that a human would follow the same semantic way: above the first block and then below the second block. It is very complicated to capture all these aspects in algorithms.

To emphasize the contrast to outdoor location-based services, the following recollects some central differences between indoor location-based services and outdoor location-based services:

Indoor location-based services need to be able to use a multitude of different methods to infer a position. The position determination can be placed either in the infrastructure or on a mobile device or even be split between both. Additionally, indoor location-based services need higher precision and reliability as compared to

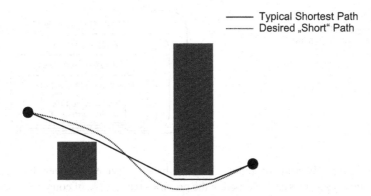

Fig. 1.4 The shortest path in free space and a equivalent path with better properties

the outdoor situation in order to correctly understand the relation between a position and the environment. With regard to the environment, indoor location-based services need a much higher diversity of information and information representations in order to provide meaning to location. Large-scale and global-scale indoor location-based services can only be realized when the complex interactions between all building owners is sufficiently organized and there is enough incentivization to support indoor location-based services. Furthermore, indoor location-based services put a challenge on the user interface as it is difficult to imagine users walking around in a building holding a smartphone in front of their face and being given audio instructions.

In comparison to other areas of location-based services, the indoor domain is special in that it has to address the following set of challenges: indoor location-based services should be based on devices and algorithms which are small, lightweight, low power, low cost, precise, ubiquitously applicable, and reliable. Though all these aspects are relevant for other location-based services, they have an even larger impact in the domain of indoor location-based services.

1.3 Indoor LBS and Ubiquitous Computing

As depicted in Fig. 1.5, indoor location-based services are a special case of location-based services. However, indoor location-based services can often not rely on location determination alone. This is due to the fact that in some areas, positioning might be too inaccurate or even completely unavailable. Therefore, the more general topics, which do not rely on location, can and must actually be integrated into many indoor location-based services in practical scenarios.

Fig. 1.5 Indoor
location-based services
embedded in more general
areas of research

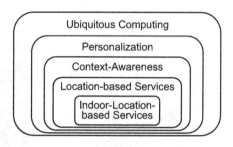

A first generalization of indoor location-based services is given by general
location-based services as described in the previous paragraph, of course. In relation
to indoor location-based services, the topic of seamless integration of indoor
and outdoor location-based services is essential for the interrelationship of these
services. Moreover, outdoor location-based services can be essential to bootstrap
some indoor location technologies with a starting position or determining the
right map to be used for particle filtering of inertial measurements when loosing
GPS while entering a building. However, for indoor location-based service, the
usefulness of outdoor location-based services is limited due to the fact that GNSS
positioning will soon be lost. In cases, where positioning is completely unavailable
(e.g., no GNSS as well as no indoor location determination), one can still continue
providing a location-based services due to context awareness. Examples are tunnels
for the outside navigation. In tunnels, the GNSS signals are typically lost. However,
navigation systems continue to give instructions. They are then either based on the
assumption that the movement speed keeps constant or by odometrical support from
the car. Inside buildings, the same can be done: one can detect the current context
of a user including its mode of transportation (walking, standing, sitting, climbing
stairs, etc.). This information can be enough to continue providing a location-based
service based on some assumptions, quite similar to the case of a tunnel for GPS-
based car navigation systems. A further generalization to location-based services is
given by personalization. Note, however, that personalization is usually defined as
modifying a service based on a user profile identified with a person. Therefore, this
is only a generalization for the special cases of indoor location-based services in
which the location information is actually identified with the user situation. Another
perspective on why personalization is not a generalization of the class of all context-
aware services is given by the observation that the broader range of personalization
does not describe the situation of the entity anymore but rather the entity itself.
Still, indoor location-based services can be enhanced and greatly improved by
incorporating elements of personalization. One prominent example is to include the
expected behavior of a person into the location-based service. In this way, a system
might completely lose track of the situation, context, and location of the mobile user
while still providing information about the status of his office and his next steps
according to the calendar application. This is the most probable incarnation of an
assumption to be made in the tunnel scenario: as long as we do not know it better, we
assume that the user is actually doing what he is most likely doing—following the

street with constant speed or walking further to the next meeting location according to the calendar application. The vision of ubiquitous computing can be seen as a global umbrella for all of these services: ubiquitous computing is achieved, when every object in the surroundings is smart and interconnected, and hence, at every time and place, context awareness and location awareness become reality.

1.4   Classical Research Areas Related to Indoor LBS

Figure 1.6 depicts a subset of classical research areas, which all integrate into indoor location-based services. A very classical area of research is geography. Geography is an integral part of all location-based services in the sense that geography studies topics such as location, mapping, and navigation. In the area of computer science, this is reflected by research in algorithmic geometry as well as in the area of geographic information systems (GIS). This is a very active research area aiming for being able to query and analyze location data efficiently. This is, of course, a central prerequisite for providing location-based services. The GIS community also defined standards such as the "OpenGIS Implementation Standard for Geographic information—Simple feature access—Part 2: SQL Option" which extends the SQL language with functions in order to process spatial datasets including geometric calculations such as point in polygon tests or the length of a "LineString." However, this extension covers only basic operations and does not suffice to provide complex location-based services. Therefore, the spatial computations involved in managing location-based services are often implemented in an ad hoc manner and optimized for the specific scenario and datasets.

Wireless communication and networks are another integral part for indoor location-based services. Not only do they provide network access and enable communication between a mobile device and some online service, they also provide

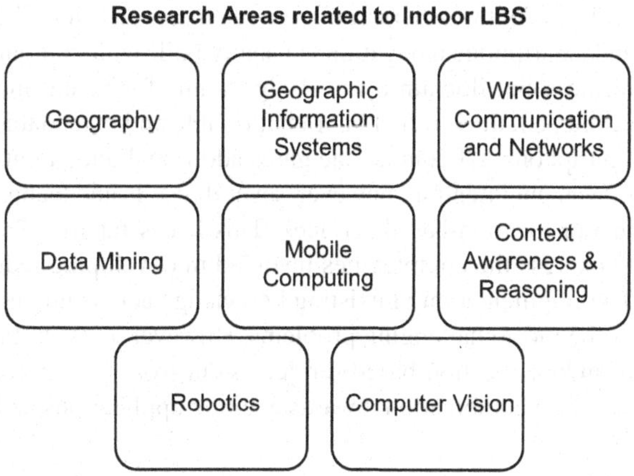

Research Areas related to Indoor LBS

Fig. 1.6 Research areas related to indoor location-based services

sources of measurable information such as signal strength, channel parameters, and timings, which are essential to indoor location estimation. The functionalities of these systems as well as the electrotechnical basis of them is a prerequisite for providing indoor location based on non-dedicated, already existing infrastructure. Data mining is added to the related research areas as the location inference problem is often and naturally expressed as a data mining problem. This is due to the fact that many positioning systems consist of two phases: an offline phase in which training data with associated location information is gathered and an online phase in which newly arriving measurements are assigned to the most probable location based on the training data from the offline phase. Moreover, the research area and developments of mobile computing are a precondition: without mobile computing devices, no computer service can be provided to which indoor location is relevant. Hence, location and especially indoor location-based services need mobile computing, and every mobile computing invention can have impact on indoor location-based services. A recent example is given by the increasing deployment of barometers. These can now be used to estimate height differences and, for example, distinguish between different floors of buildings. The area of context-awareness is tightly coupled with semantic reasoning techniques and is an additional topic of influence for location-based services. For proactive location-based services, the intention of the user might be detected based on context information, and the transformation of this information into a navigational task could be performed by reasoning. Moreover, indoor location-based services are instances of context-aware services, and hence, the contribution is in both directions. The field of robotics is one of the older and established fields in which topics such as mapping and navigation have been researched. In the domain of autonomous robotics, robots should be able to orient themselves even in previously unknown environments and often inside buildings. The central difference to indoor location-based services is given by two common assumptions about robots: the computer controlling the robot controls the movement of the robot. This amounts to the availability of the actual intention of the robot (how did he try to move) to the positioning algorithms. Furthermore, a robot is equipped with dedicated, accurate, and possibly expensive infrastructure such as depth cameras or laser scanners. A bridge between this robotic domain and smartphone navigation is usually built with computer vision techniques. While in the robot domain a lot of algorithms for visual mapping and navigation have developed, these might be transformable and extendable to human beings holding a smartphone. Of course, one gains additional uncertainty by not knowing the actual movement intention; however, several results and techniques can be transformed for indoor location-based services. This list is far from being complete and intended to express the complexities involved in developing indoor location-based services. The fragmentation of existing knowledge across numerous isolated domains is one of the most challenging problems. However, most of the fascination and aesthetics of indoor location-based services stem from the interconnection of these domains and their combination towards exciting applications and results.

1.5   Classical Applications of LBS

There is a vast amount of possible location-based services most of which are also possible inside buildings. This section provides a short list of very common applications each illustrated with a concrete example and a list of prerequisites for this type of service. In this way, this section motivates the many different problems covered in this book by an intuitive and clear application scenario.

1.5.1   Information Services

Information services comprise services, which select, filter, and rate a set of information in relation to the location of some targets, possibly the user in personal location-based services or even some other entity in fleet management.

Example 1.1 (Smart Logistics) A logistics system is using an external online service to gather traffic information. However, this traffic information is available only using a location-based query interface to extract information relevant to some area. This traffic information service produces cost and bills the service consumer an amount of money related to the amount of traffic information being requested. The logistics system exploits the available GPS information as well as the planned routes for these trucks in order to select the minimum amount of traffic information in order to optimize the process.

In this example, the location-based service relies on valuable information, and positioning is being used to extract only the relevant subset of the available information in order to reduce cost, bandwidth consumption, and calculations.

Example 1.2 (Map Application) A smartphone provides the user with a map of his surroundings. Therefore, the smartphone tries to detect the current position of the user, retrieves a map from some online database, and depicts this map and the estimated location of the user inside this map on the screen. The user is able to scroll and zoom the screen in order to orient himself.

In this example, a user is enabled to orient in unknown environment by providing a map of the surroundings. Interestingly, the system is set up in a way such that only the relevant parts of possibly very large maps (e.g., containing the complete world) can be provided to the mobile user over the Internet. This makes the service scalable to all areas where Internet access is given and map information is available.

Example 1.3 (Bar Notification) A user has installed a smartphone app which proactively informs him about the best bars around. While the user is moving through an unknown city, the phone rings and tells him that he is very near to this or that bar and that he might check out this bar getting a discount for the first drink.

Example 1.4 (Proactive Tourist Information) A tourist uses an app that is tailored to providing information about city tourists. Then, while the tourist moves around a city, the smartphone informs him every time he comes near to a place of interest including all major touristic sites. Additionally, the application provides background information about the current location such as the year of construction of buildings.

In this example, the smartphone needs the possibility to locate himself, and the service needs a mechanism to efficiently track the k next neighbors of the current position from a table of relevant locations in real time.

Prerequisites: Localization, Map Information, Mobile Internet, Interesting Points, Spatial Algorithms

1.5.2   Navigation Service

Example 1.5 (Indoor Navigation) A mobile device provides a map service in which a map of the surroundings of the user is provided together with an indication of the actual location of the mobile device in relation to this map. Furthermore, the user is able to input a description of his destination, and the service provides the shortest path towards the destination, depicts this path on the map, and provides turn-by-turn guidance to the mobile user.

This example gives the primary motivation of indoor location-based services as taken from the outside world. Users shall be able to navigate inside buildings as easy as they can navigate on streets using a vehicle navigation system and GPS.

Example 1.6 (Personalized Interactive Maps) A wheelchair user often has additional difficulties in navigating through complex public buildings as the signage is optimized for non-wheelchair users, and he suddenly gets stuck near some staircase or similar. In order to address this special group of building users without additional signage, a building operator provides an interactive map of a building in which the user can select that he needs to take elevators in order to change floors and in which the best route using these constraints is planned and shown.

This example of an indoor location-based service makes clear that a location-based service is possible even without a computer system which calculates location. This task can be left to the user who has to navigate to the correct location using the service user interface.

Prerequisites: Localization, Map Information, Mobile Internet, Interesting Points, Routing, Tracking, Guidance

1.5.3 Safety-of-Life Applications

Safety-of-life applications provide a very relevant area of applications for several reasons: First of all, funding research is more easy for application areas in line with public interests. Additionally, it is feasible to use expensive elements inside the navigation service as the number of objects to be tracked is limited by the number of emergency workers involved. This makes some advanced hardware feasible, such as when navigating firefighters.

Example 1.7 (E-911) Every time that a mobile user calls the emergency number, the cellular network infrastructure determines the location of the caller with a previously defined accuracy and provides this location information to rescue workers, which have less problems to find the actual emergency spot in crowded or cluttered environments.

Example 1.8 (Ambient Assisted Living) A computer system passively tracks the movements of elder people inside their own homes. Thereby, significant deviations from the daily routine can be detected, and other family members or doctors can be informed accordingly.
Prerequisites: Localization, Pattern Recognition

1.5.4 Retail and Commerce

Retail and commerce are interesting application areas for location-based services as they often do not need unrealistic accuracy in location information in order to work. Thereby, a lot of examples can be set up in a rather simple way including proximity localization.

Example 1.9 (Proximity Marketing) A shop provides discounts to all users carrying a smartphone and passing by the shop at a specific time. Therefore, the shop provides some infrastructure, and the smartphones have a software that is capable to detect this infrastructure and provide the discount to the user.

This type of application examples has become very prominent in the last years due to Apple investing into Bluetooth Low Energy (BLE), iBeacon, and Passbook. These three elements comprise a complete ecosystem for services comparable to the example for Apple devices working inside buildings.

Example 1.10 (Behavioral Analysis) A shop provides a location-based service which provides discounts based on the proximity to some shop. In the background, all location data is collected and being used in order to perform behavioral analysis of individual customers in order to target them according to their preferences.

From a technical perspective, this example is quite similar to the previous example of proximity marketing. However, it reveals the central problem for location-based service users: privacy is difficult to guarantee, and therefore, the exploitation of the collected data for behavioral analysis or similar studies is to be expected.

Prerequisites: Proximity Detection, Mobile Computing, Privacy

1.5.5   Management

Example 1.11 (Process Observation) A car manufacturer uses a checklist during final quality checks of a car. However, after a long working day, workers inadvertently forget steps. A high-accuracy location determination system tracks all actions actually performed by the worker and automatically remind the worker of the forgotten steps inside the quality control process.

Prerequisites: Accurate Localization, Activity Recognition, Tracking, Process Modeling

1.5.6   Social Networking and Joint Activities

Example 1.12 (Buddy Finder) A social networking application continuously tracks the k nearest friends of some user with respect to the current location of the user and provides them to the user for messaging.

Example 1.13 (Joint Activities) A social network service detects a situation in which at least three social network users with similar interests in some activity such as sports or making music come near each other and have marked themselves as interested in performing this joint activity. In the situation, all users are informed about the situation and can communicate with each other in order to group up for performing the joint activity.

These examples are very simple services with respect to the requirements on localization technology. Therefore, they can easily be transformed to the inner of buildings as proximity detection is not that hard. However, it involves complicated spatial computations and provide examples for the difficulty of providing privacy in location-based services.

Prerequisites: Localization, Clique Detection, Mobile Internet

1.5.7   Gaming

A very classical location-based board game is called Scotland Yard and was classically played on a simplified map of London. This classical board game is easily and often transformed into a location-based game smartphone game:

Example 1.14 (Scotland Yard) One player, called Mr. X, is moving around the city and shows his location only at stated intervals. The other player locations are always visible to all players including Mr. X. The task for Mr. X is now to move around the map with different and secret modes of transportation without another player coming to the same spot of the map.

Prerequisites: Localization, Map Information, Tracking, Proximity Detection, Mobile Internet

1.6   A Short History of Navigation

Indoor location-based services seem to be a reincarnation of many classical problems in the history of navigation. For each of the following milestones for nautical navigation, we will indicate where this very same problem is challenging for indoor location-based services.

The first problem hindering navigation was the problem of knowing the location of a ship on the high sea. For a very long time, this was impossible, and ships were always traveling in sight of the shore.

This movement in sight of the shore translates to the area in which location determination technology is available inside buildings.

Around 1740 several constructions equivalent to the so-called sextant were known. The sextant is a device with which the angle between two objects can be measured. It was typically used between stars or the sun and the horizon. The classical sextant constructions reached an accuracy of an arc minute, which amounts to a positioning error of approximately a nautical mile. Interestingly, the widespread use of the sextant was replaced only recently by the global availability of GPS. This is due to the fact that the sextant provided global positioning with quite high accuracy.

A technique comparable to the sextant, which is able to find an accurate location merely everywhere in the world and completely independent from location-based infrastructure, is still missing for indoor location-based services.

However, the central precondition on calculating location from angles is a very exact time. This is quite clear when thinking about the moon or the sun as the heavenly bodies used for measuring location. Over time, the angle between the horizon and these objects is changing even when location is kept constant. This problem of tracking the time on a ship with sufficient accuracy has led to the invention of the chronometer. The chronometer is a clock, which provides very exact time even on the high seas. This was a very challenging problem for quite a long time as mechanical clocks were sensitive to the movements of a ship. At the beginning of the seventeenth century, Harrison developed the first chronometer, which would have had sufficient accuracy. However, it was too expensive and it took quite a while until his construction has been made affordable. These chronometers where widely applied in the seafaring until the invention of quartz clocks, which provide better accuracy at lower cost and size.

> The problem of knowing an exact time is an inherent challenge for indoor location determination. Many positioning approaches rely on measuring delays and perturbations of signals traveling at the speed of light. These effects can only be observed with very high-accuracy clocks and distributed clock synchronization.

Parallel to the invention of different techniques to find the actual location of a ship on the high seas, it became possible to create and refine maps of the surroundings over time. Due to some unclarity of how to exactly define which visual representation of some location information constitutes a "map," research is still discussing which were the first maps in the history of humankind. The first world maps are assumed to be from the Babylonians from the ninth century BC. Over time, the observation of different people was integrated step by step into world maps by copying from different sources and adding some details from the own observations of some cartographer. However, it was impossible until the invention of the sextant to accurately modify, extend, and correct maps, and many small and highly accurate maps exist which do not fit together into a global consistent map. With the sextant, one was able to find out some location accurately in two different maps and, therefore, integrate map information. Since the middle of the twentieth century, aerial photography, satellite imagery, and remote sensing together with the invention of computers provide efficient, scalable, and precise methods for creating maps of the world.

> For indoor location-based services, we are somehow in the process where a multitude of different, highly accurate, local maps exist such as before the sextant. But we are still unable to integrate and correct these datasets due to the unavailability of a global reference system in which accurate indoor position determination is possible.

1.7   Structure of This Book

The structure of this book is oriented towards the progress of learning. Readers, who are not familiar with different topics, should be able to follow the book from the beginning towards the end. Moreover, chapters have been written in isolation with each other though some cross-references are given. This is to address people who are familiar with a subset of the topics but curious towards others. In some cases, I decided to include references to later chapters in order to make a lookup of information more efficient for experienced readers. Consequently, all chapters are basically self-contained, of course expecting some rough knowledge about the contents of preceding chapters. This implies some redundancy between chapters of this book.

Chapter 2 provides details about *Wireless Communication Systems* as a lot of positioning approaches, both indoors and outdoors, are based on these techniques. Moreover, some different *Sensor* hardware is explained. This is done on a simplified level intended to let readers get a feeling on which properties should be expected for the different sensor sources.

The following Chap. 3 explains the mathematical and algorithmic foundations of *Basic Positioning Techniques* based on measurements including the underlying geometric principles as well as the observable variables including timing, time difference, angle of arrival, angle of emission, signal strength, acceleration, rotation, and more. The following Chap. 4 on *Building Modeling* reflects the need of indoor positioning systems to integrate environmental information. The chapter collects the needed information and algorithms from the domain of geometric information systems (GIS), algorithms, coordinate systems, geometric markup language (GML), City GML, and similar approaches.

Having environmental models and basic positioning techniques in place, the area of *Position Refinement* is presented in Chap. 5. There are many possible methods to improve the position information reported by basic positioning techniques, often by taking the evolution of position over time and the relation of positions with respect to map information into account. This chapter includes techniques such as recursive least squares method, Kalman filtering, and particle filtering.

Chapter 6 on *Trajectory Computing* provides a more general point of view as compared to the previous chapter. As the position refinement methodology showed, the time domain can add valuable information to positioning systems. This chapter is based on generalizing positioning systems to use location trajectories (sequences of locations) as well as measurement trajectories (sequences of measurements). This chapter provides basic and advanced algorithms and results from this domain and explains their impact on indoor location-based services, which might not rely on the position of a user alone.

While outdoors the position as well as a trajectory is often enough to determine most location events (corners, turns, arrivals at points of interest, etc.), inside buildings much more events can become relevant, and many applications have to rely on a different technology to find these events. Hence, Chap. 7 provides a hint on general *Event Detection for Indoor LBS*. This can, again, be seen as a generalization of the previous chapter in which trajectories are usually dense in time. In contrast to that, events will be sparse in time and reported only when they occur. One of the most prevalent examples is that the performance of Wi-Fi localization is pretty good in two-dimensional space but can have difficulties to determine the right floor and detect floor changes. These systems can be augmented, for example, by using a barometer indicating floor changes in a reliable way.

The next Chap. 8 covers *Simultaneous Localization and Mapping in Buildings*. Simultaneous Localization and Mapping (SLAM) refers to a family of algorithms to generate map information while positioning the moving target at the same time. SLAM algorithms rely on many elementary algorithms such as feature point tracking, loop detection, Random Sample Consensus, and more. This chapter aims to bring together a closed exposition of these algorithms and their application inside buildings.

In Chap. 9, classical definitions of anonymity and privacy are being introduced including k-anonymity, l-diversity, and ϵ-differential privacy. Furthermore, multiparty computation as well as cryptographic private information retrieval is introduced and discussed.

The final Chap. 10 follows the structure of the book and highlights some areas where additional research and development might take place. This can be seen as a reflection of what has actually been achieved and what is missing and, therefore, provides a conclusion to the book.

1.8   Further Reading

This book intends to bring together information on indoor location-based services from various domains. Therefore, a lot of textbooks exist, which cover aspects of this book in greater detail or in a different context. In this section, I want to highlight some of them as they might be relevant to the reader: For the topic of positioning and navigation, there are several useful classical books including Hoffman-Wellenhof, Legat, and Wieser's *Navigation* [4] and for a very detailed and up-to-date exposition including some implementation examples on DVD the book of Paul D. Groves *Principles of GNSS, Inertial, and Multisensor Integrated Navigation Systems* [3]. Another book with a focus on concretely available indoor location systems is given by Subrata Goswami *Indoor Location Technologies* [2]. For general location-based services, the book of Axel Küpper *Location-Based Services: Fundamentals and Operation* provides additional material on communication and management of location-based services [5]. For the area of computational geometry, I want to recommend the book of deBerg, Cheong, Kreveld, and Overmars *Computation Geometry: Algorithms and Applications* [1]. Furthermore, a good collection of articles on trajectory computing has been published in Zheng and Zhou (eds.): *Computing with Spatial Trajectories* [7]. This list of books is my personal selection; there are lots of additional sources. However, these books have been most influential to me.

References

1. De Berg, M., Van Kreveld, M., Overmars, M., Schwarzkopf, O.C.: Computational Geometry. Springer, Berlin/Heidelberg (2000)
2. Goswami, S.: Indoor Location Technologies. Springer, New York (2013)

3. Groves, P.D.: Principles of GNSS, Inertial, and Multisensor Integrated Navigation Systems. Artech House, London (2013)
4. Hofmann-Wellenhof, B., Legat, K., Wieser, M.: Navigation – Principles of Positioning and Guidance. Springer, Wien (2003)
5. Küpper, A.: Location-Based Services: Fundamentals and Operation. Chichester, West Sussex, England, Wiley (2005)
6. Virrantaus, K., Markkula, J., Garmash, A., Terziyan, V., Veijalainen, J., Katanosov, A., Tirri, H.: Developing gis-supported location-based services. In: Proceedings of the Second International Conference on Web Information Systems Engineering, 2001, vol. 2, pp. 66–75. IEEE, Kyoto, Japan (2001)
7. Zheng, Y., Zhou, X.: Computing with Spatial Trajectories. Springer, New York (2011)

Chapter 2
Prerequisites

A little inaccuracy sometimes saves tons of explanation.

Saki

Indoor location-based services become possible due to two important developments in computing: *mobile computing and mobile communication systems* as well as *sensor technology* in mobile devices. From the beginning on, mobile computing devices had some systems for communication like infrared or cable ports in order to synchronize working data with personal computers. In the last decade, mobile Internet access using Wi-Fi and cellular networks has made these interfaces obsolete for many mobile communication devices and enabled the invention of smartphones which are able to work on remote data over the Internet efficiently. Smartphones and other small devices make indoor location-based services more relevant as these devices are often used in a highly mobile context. This chapter introduces the technical developments which made ubiquitous network access via GSM, LTE, or Wi-Fi possible. Understanding mobile communication systems is also essential for indoor location-based services from another perspective: wireless communication infrastructures serve as a welcome source of location-dependent information based on signal strength, identifications, or their inherent location management for link optimization and handover.

The second aspect enabling indoor location-based services consists of discussing the sensor systems typically available to a modern mobile device. For indoor location-based services, no single sensor system suffices to reach a good level of location awareness. Therefore, the advantageous integration of those many available sensor systems leads to successful indoor localization. Therefore, an introduction to their working and measurement principles is given.

2.1 Mobile Computing and Mobile Communication

Mobile communication consists of two aspects: mobility and communication. Mobility deals with the implications of movement and the introduction of mobile computers such as laptops, PDAs, or smartphones. Communication deals with information exchange between mobile computing units. Classical computer systems

© Springer International Publishing Switzerland 2014
M. Werner, *Indoor Location-Based Services*, DOI 10.1007/978-3-319-10699-1_2

have been fixed to a specific location. The reasons were size and electrical power supply. Those fixed computers can have access to some network or the Internet, typically bound by a cable. The next step in the evolution of mobile computing was the creation of laptops, which are able to work without a power supply using a battery. Laptops typically have many interfaces for interconnection with other devices such as Ethernet, modem, Bluetooth, or Wi-Fi. The next step in evolution was the creation of PDAs such as the Palm providing a small handheld device allowing a limited set of functions to be used mobile. These devices were typically connected to a personal computer for synchronization of information. The last step in evolution is the creation of smartphones, which bring together functionalities of personal computers, PDAs, and cellular phones providing Internet browsing capabilities, email software, social networks, and a complete computing platform comparable to modern personal computers.

2.1.1 Mobility

From one perspective, mobility can be defined to be the ability of a specific entity to change location. This perspective directly gives rise to the following classes of important types of mobility, which differ only with respect to which entity is mobile:

- Terminal mobility
- Personal mobility
- Service mobility
- Session mobility
- Data mobility
- Code mobility

Terminal mobility is used for systems, which allow a special entity called terminal to change location. A classical example is a cellular phone which is able to move between adjacent cells without service interruption and even without user attention. Similarly, a GPS receiver belongs to the class of devices with terminal mobility, as it can freely move around the world while in use. Typical services enabling terminal mobility range from global communication systems as in the case of GPS over systems implementing handover, where the point of attachment to some network is updated, to paging and push services, where a specific terminal is localized inside a network to provide information such as incoming calls or short messages from the short message service (SMS) to this device.

A system provides personal mobility, if the system is personalized, and this personalization can be moved between different hardware or locations. A classical example is the subscriber identity module (SIM) card, which is a small device introduced in the GSM cellular network allowing the change of phone hardware by holding network and subscriber information as well as contact information. Classical services in the area of personal mobility are call forwarding, where a service is following a specific user to another phone number, and roaming, where a specific user is allowed to move to a foreign network.

Service mobility is the property that a specific service can be used from different locations in the same manner. Classical examples are web services such as online banking or webmail. In these cases, the service is defined somewhere on the Internet and can be delivered to any device that has an Internet connection and a browser. A highly dynamic form of service mobility is given, when the service is adaptive or context-aware. These services change behavior in a way that helps users in a specific situation known to the service. A simple example is content adaption, when a webmail platform provides low-resolution pictures to mobile handheld devices as a preview.

A system supports session mobility, when a session can be interrupted and move to a new location. This means that a specific session, its properties, and associated resources can be moved to another place, e.g., from a personal computer to a handheld device. Real-world examples are still missing for the following reasons: session mobility is very complex and is only applicable to systems, which have suitable content adaption. Moving a high-resolution video session to a smartphone is still impossible.

Data mobility is the ability of several systems to exchange service data. Data mobility is often achieved via standardization of file formats or interfaces. In these cases, the systems can be coupled by exchanging data via interfaces. However, data mobility poses difficulties for real-time systems and in cases where the representation of the data has to be different for both systems. In these cases, the data transformations can quickly become too complex or the amount of traffic generated by data mobility cannot be handled efficiently.

Code mobility is an orthogonal approach to data mobility. It is given by making the executable code mobile and using communication technologies in order to move it towards the data. A simple real-world example is given by using JavaScript or Flash inside a webpage. While the webpage contains some data, the intended visualization by means of animation of this data can be quite complex. In these cases, the animations might be programmed instead of transmitted as data thereby reducing the amount of data that needs to be transferred. Another situation is given, when the data is very sensitive and should not be communicated at all. In this situation, a program can be moved towards the data for processing.

All types of mobility induce security and privacy risks due to mobility. In all cases, an actual change of location needs to be proven, and the processing of a change of location and behavior of a specific entity needs to be authorized. Also in most cases, some data about the mobility operation is going to be stored somewhere, for example, in the logs of cellular network base stations. As this information is usually collected transparent to the user, he will usually not be aware of that it happens and cannot assess the privacy and security risk. As an example, cellular network provides usually collect enough information somewhere in their distributed network to infer past location data and daily routines of a specific person using the network.

Another widely used classification of mobility is given by the scale of movement. It is customary to consider the following three scales:

- Micro mobility,
- Macro mobility, and
- Global mobility.

For computer networks, the first two classes can be easily mapped to different OSI layers based on where the mobility functions are implemented:

Micro mobility means management of mobility inside a given network. Mobility management is taken care of at OSI layer 2 such that higher layer network addresses do not change. A simple example is given by Wi-Fi-Extended Service Sets (ESS). In this situation, the Wi-Fi network is identified by a network name, and clients connect to the best access point providing access to a network of this name in their surroundings. The best access point can, for example, be chosen based on the received signal strength (RSS). If a mobile terminal moves around, the "best" access point will change, and the Wi-Fi system will transparently associate with a different access point, without changing the IP address or any other higher layer information.

Macro mobility (intra-domain mobility) is typically given inside larger organizations, where the movement is not limited to a single access network technology such as Wi-Fi. In this situation, mobility management can be done on the network layer (OSI layer 3), such that a change of network address does not affect higher layers. The most influential protocol establishing macro mobility is Mobile IP. In Mobile IP, each device has a fixed home address, and an entity called home agent is taking care of the dynamic address of the device in a foreign network and dynamically routes traffic between the mobile device and its home address.

The term global mobility (inter-domain mobility) is reserved for mobility management at higher layers. In this area, there are "over-the-top" mobility services, which provide access to a virtual network on top of the real network, where the virtual network address (e.g., IP address) can easily be kept static. Global mobility is also much a topic of international and intercultural organization, communication, and standardization. An example of a protocol establishing global mobility is GSM, where users can access the same services from numerous countries. Figure 2.1 depicts the explained mobility scales.

A third perspective on mobility is given by the level of mobility support, which a specific system, device, or terminal can give. One distinguishes between

- nomadic systems,
- portable systems,
- limited mobility systems, and
- real-time mobility systems.

Nomadic systems can freely change location but become stationary during communication. Communication can be interrupted by movement. A classical example of a nomadic system is mobile satellite receivers, where the antenna has to be oriented towards a specific satellite, before receiving a signal. With respect to indoor location-based services, some visual communication systems are nomadic

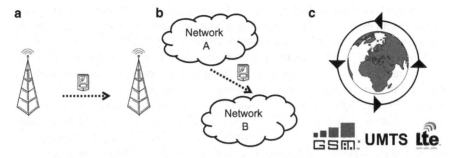

Fig. 2.1 Different scales of mobility. (**a**) Micro mobility. (**b**) Macro mobility. (**c**) Global mobility

systems. In order to scan a QR code, for example, the phone camera has to be pointed onto the QR code, and the user has to wait for a sufficiently sharp image. A constantly moving camera is unable to decode QR codes in many situations due to motion blur.

Portable systems keep basic connectivity to some access network, but movement speed or movement area is limited, and short service interruptions are to be expected.

Systems with limited mobility support provide continuous network access for non-real-time applications. An example is given by cellular networks: Though transparent handover between cells is implemented, delays and a small number of actual disconnections are allowed to occur during handover.

Systems with real-time mobility support provide mechanisms mobility management such as handover, which are reliable and fast enough to provide connectivity to real-time applications such as telephony and real-time multimedia streaming.

In summary, mobility can be viewed from at least three perspectives: firstly based on which entity is mobile, secondly based on the scale of mobility, and finally based on the amount of mobility support. In general, mobility is one of the central elements in location-based services: without mobility, location-based services are neither needed nor possible. However, computer services are based on communication, and therefore, the next section subsumes communication systems in general as a preparation for the discussion of real-world mobile communication systems.

2.1.2 Communication Systems

Communication, in general, is the process of encoding and transmitting some information between entities. We know many forms of communication, for example, speech, visual symbolism, signals, writing, and behavior. All these types of communication can be seen as variants of the same general methodology.

Therefore, communication systems refer to systems, which are in compliance with a system model. An early model for a communication system was given in [9] consisting of five entities interconnected as depicted in Fig. 2.2.

With respect to this model, the main components of a communication system are an information source, a transmitter, a noise source, a receiver, and a destination. An information source generates a message or a sequence of messages to be transmitted. Messages are basically considered as continuous functions in several variables, often including a time variable. A simple example would be a voice signal encoded in PCM, that is, a sampling of a continuous function of time given by the voltage level of a recording microphone. Once the information source has generated a message, this message is given to the transmitter, which translates this message into a form suitable for transmission over a medium. This form of the message is called signal. During transmission, effects on the medium such as noise or superposition in a shared medium will alter the signal resulting in a received signal at the receiver. The receiver basically performs the inverse operation of the transmitter translating the received signal back into a message. This message is then given to the destination, and the communication process is complete.

The fundamental problem of communication is the approximate reconstruction of a message from the received signal in cases where differences occur between the signal and the received signal.

Example 2.1 As an example, consider a written letter sent via mail. The entities of the communication system in this case could be given as follows:

- *Information Source:* The author of the letter.
- *Transmitter:* The person writing down the letter with a pen and a sheet of paper.
- *Signal:* The sheet of paper containing handwriting.
- *Noise:* Possibly blur of ink due to contact with humidity.
- *Received Signal:* The blurred sheet of paper containing unclear handwriting.
- *Destination:* The recipient of the letter hopefully getting the same message as intended by the information source.

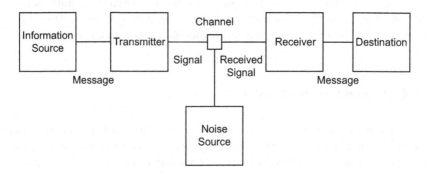

Fig. 2.2 Shannon model of a communication system [9]

For written language, there is a high degree of redundancy in human languages. Think of the simple rule "i before e except after c," which imposes a redundancy relation between letters i and e in English language. This redundancy enables a reader to deal with corrupted words. This redundancy has also been successfully applied in the field of cryptanalysis for ciphertext-only attacks using, for example, the index-of-coincidence technique.

This type of redundancy in natural languages is of general interest. It is possible to design communication systems, which make use of this redundancy, and it is possible to reuse this type of redundancy to generate "realistic" messages for the evaluation of communication systems and calculate upper bounds on the efficiency of such a system.

There are mainly two types of statistical language approximations using either the character structure or the word structure of a language. For characters, we can estimate the relative frequencies of different characters in natural language. Therefore, we need a representative set of text and count frequencies. If the text basis is large enough, these relative frequencies will approach the probability of a character appearing in a character sequence in the given language. This defines the *first-order* approximation to the language. Of course, the choice of adjacent characters is not independent in most languages. Hence, estimating relative frequencies of two-character sequences gives a better approximation to the original language, called *second-order* approximation. This scheme of generating approximations by considering sequences of adjacencies of letters can of course be continued, but instead using the character structure, it is better, according to [9], to switch to the natural word structure. Again, words (or sequences of two, three, or more words) in a large text database are counted, and relative frequencies are used to estimate the occurrence probability of them. These probability distributions can then be used to generate random, yet realistic, sequences of text, which follow the statistical properties of the base text up to the given order.

Example 2.2 Using the full text of the famous work Oliver Twist written by Charles Dickens, we can calculate the following letter probabilities (limiting to lower case characters including space). The file used here consists of 846.921 characters with individual frequencies tabulated in Table 2.1, where also relative frequencies are given. For example the space character appeared 149.151 times, that is with a relative frequency of $149.151/846.921 \approx 0.176$, which means, that in average, every fifth to sixth letter in this work is a space.

Turning this observation into a stochastic process, independently choosing characters with probabilities as given in the table, leads to text sequences, which are well designed to test a natural language communication system. Table 2.2 gives some examples of different word-level approximations to the language as used in Dickens work.

The redundancy of text is linked with a very fundamental topic of computer science, the measurement of information. For a communication systems, a fundamental question is, how much information is generated by an information source. Think of a simple information source given by a stochastic process, which chooses with probability one a specific letter, say, "a," and with probability zero all other letters. How much information is involved in choosing the second letter of a sequence generated by this stochastic information source? None. If, however, all letters are chosen with the same probability $1/n$, n being the size of the alphabet, we would expect a maximal value for an information measure, as there is much choice involved in choosing one out of the n letters and there is no information on which letter would be a good guess.

Table 2.1 Table defining the first-order approximation to Oliver Twist

Char.	Count	Relative count	Char.	Count	Relative count
Space	149,151	0.17611	n	47,316	0.05587
a	54,245	0.06405	o	53,838	0.06357
b	11,363	0.01342	p	12,451	0.01470
c	16,449	0.01942	q	696	0.00082
d	32,908	0.03886	r	41,701	0.04924
e	87,225	0.10299	s	41,583	0.04910
f	14,760	0.01743	t	61,607	0.07274
g	15,283	0.01805	u	19,103	0.02256
h	45,727	0.05399	v	6,625	0.00782
i	48,213	0.05693	w	17,081	0.02017
j	1,023	0.00121	x	955	0.00113
k	6,293	0.00743	y	14,453	0.01707
l	28,339	0.03346	z	161	0.00019
m	18,372	0.02169			

Table 2.2 Some word-level approximations to language as a stochastic process

Approximation	Typical example
First-order word	it's time lessons in the robber caught reached post-chaise for
Second-order word	you couldn't was the face in to death
Third-order word	boy, named oliver mug of ale, and tells him that,where they're demanded monks, sternly
Fourth-order word	and a very good and shook him cordially the mysterious youth
Ninth-order word	I don't think that would answer my purpose. Ain't this circumstance, added to the length of his legs, was too much

More generally, let S be a finite set of n events and let each element $s_i \in S$ be assigned a probability p_i of occurrence, such that $\sum_{i=1}^{n} p_i = 1$. If there is a measure $H(p_1, \ldots, p_n)$, which gives the amount of information produced by an occurrence of one element out of S using the stochastic process defined by the probabilities, one should expect at least the following properties of the measure H:

1. H is continuous in p_i.
2. If all p_i are equal, that is, $p_i = \frac{1}{n}$, then H is monotonous function of n.

3. If a choice is broken down into two successive choices, then H decomposes as a weighted sum:

$$H(a,b,c) = H(a,b+c) + (b+c)H\left(\frac{b}{b+c}, \frac{c}{b+c}\right) \text{ ,where } a+b+c = 1$$

These three properties can be motivated as follows: the first property, namely that H is continuous in p_i, means approximately that when the probabilities change by a small amount, then the information measure changes also only by a small amount. This is to be expected as the stochastic process generating sequences is also changed only in those seldom cases in which the small change of letter probabilities leads to a different letter being chosen.

The second property can be rephrased as follows: for equally probable letters, the amount of information increases with the number of different letters. A choice of a letter from a one-element set does not contain any information, while choosing an element out of a large set needs a lot of information.

The third property can also be rephrased into natural language as follows: choosing one out of three letters A, B, and C (e.g., $H(a,b,c)$, where a, b, and c are the relative probabilities of A, B, and C, respectively) needs the same information as choosing between A and "B or C" followed by choosing between B and C, when "B or C" was chosen. The first choice needs information $H(a, b + c)$, and the second choice needs information $H\left(\frac{b}{b+c}, \frac{c}{b+c}\right)$. Furthermore, the second choice, and therefore the information for the second choice, is only needed in cases, where A has not been chosen, that is, with a relative fraction of $(b + c)$. This explains the factor $(b + c)$ in front of the second summand.

Having motivated three properties for an information measure, we are finished due to the following beautiful fact, often called the foundation of information theory: these three properties define an information measure uniquely up to a scalar related to the choice of a unit of information.

Theorem 2.1 *A function H with properties as above exists and is of the form*

$$H = -K \sum_{i=1}^{n} p_i \log p_i,$$

where K is a positive number given by the choice of a unit of information. For the information unit binary digit *(bit), K is given by $\frac{1}{\log 2}$. The form H is called entropy.*

The factor $K = \frac{1}{\log 2}$ given above changes the natural logarithm $\log$ with base e into the logarithm with respect to the base 2, suitable for using *bits*. If the unit of information shall be a digit including zero, then the factor K would be $K = \frac{1}{\log 10}$, and the information unit *digit* would contain $\frac{\log 10}{\log 2} \approx 3\frac{1}{3}$ bit of information.

In the literature, sometimes the factor K is given implicitly by specifying a different base for the logarithm as, for example, in

$$H = -\sum_{i=1}^{n} p_i \log_2 p_i.$$

For a probability value p ranging from 0 to 1, the entropy $H(p, 1-p)$ is depicted in Fig. 2.3.

Using only the information contained in the first-order approximation of the example, the entropy can be readily computed and amounts to 4.1 bit per letter. This is a very high value as the actual information for English language is estimated to around 1 bit per letter; however, this is the entropy considering only the first-order letter structure of the English language and, hence, does not reflect the actual entropy, which is lower due to redundancy relations between letters and due to the fact that only very few sequences generated according to the first-order letter structure are English words.

Some calculations for the English language give an information value of 1 bit per letter [8, p. 234], meaning that in theory a compression scheme exists, which shrinks English ASCII text by a factor of 8. The concept of Shannon entropy was introduced together with the Shannon communication model in [9], where also some techniques for the computation of this entropy are given.

The concept of entropy is also used in data mining applications, especially decision tree induction. For example, the ID3 and the refined C4.5 algorithm applied to discrete data [5, 6] calculate a measure of *information gain*, which estimates how much information about the dataset is gained in a specific split. The information gain is the difference of information values, which are essentially given by the entropy of the relative frequencies of attributes in the branches of the candidate tree. For details, see [3, Ch. 4.3].

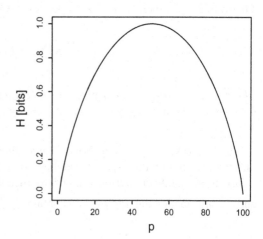

Fig. 2.3 The entropy of one probability p and its opposite $1-p$

2.2 Wireless Communication

Wireless communication systems are among the most important inventions for providing indoor location-based services. A location-based service needs to be mobile, e.g., be used from different locations. Furthermore, a location-based service is often provided based on communication between the mobile entity and some infrastructure such as the Internet.

Wireless communication systems provide both: they are often used to transmit information between the mobile device running a location-based service and a service backend, and they can be used to infer the location of a mobile device. This section, therefore, recollects some basic properties and techniques from the area of wireless communication systems.

2.2.1 Signals

As defined in the Shannon model of communication, signals are encodings of information suitable for propagation through a medium. For the purpose of wireless communication, it is best to think of signals as functions of time. If this function is continuous, the signal is called *analog*.

$$s(t) \rightarrow \mathbb{R}$$

A typical example for an analog signal is the voltage level of a microphone attached to a battery. If the signal function takes discrete values, it is called *non-analog*. The most common special case is a *digital* signal, which is a signal with exactly two states, i.e., a function

$$s(t) \rightarrow \{0, 1\}.$$

2.2.1.1 Electromagnetic Waves

An *electromagnetic wave* consists of an oscillating electrical and magnetic field. The magnetic field is usually orthogonal to the electrical field, and those are both orthogonal on the direction of propagation. Figure 2.4 shows an example.

Electromagnetic waves propagate through vacuum or media (e.g., air, water, etc.). The *propagation speed* is very high, approximately 300,000 km/s for vacuum, slightly slower in different media. Every oscillating electric charge emits electromagnetic waves, which can only be observed by their own influence on movable electric charge.

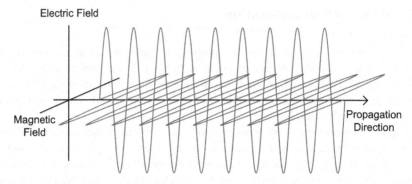

Fig. 2.4 An electromagnetic wave. The electrical field is depicted in vertical direction, the magnetic field in horizontal direction, the wave is traveling from left to right

At a specific point in time and a specific location, there are usually many electromagnetic waves in superposition. A suitable and for our purposes sufficient mathematical model for electromagnetic signals is given by

$$s(t) = \sum_{i=0}^{\infty} A_i \sin(2\pi f_i t + \phi_i). \tag{2.1}$$

In this representation, an electromagnetic wave is transformed to a voltage-level signal by an antenna and can be given as the superposition (e.g., addition) of elementary waves of the form

$$A \sin(2\pi f t + \phi)$$

with constant parameters

- Amplitude $A \geq 0$,
- Frequency $f \geq 0$, and
- Phase ϕ.

There is another important measure related to the frequency and propagation speed called wavelength. The wavelength λ of an elementary wave $s(t)$ is defined to be the minimal length of periodicity, i.e., $s(t) = s(t + \lambda)$ for all t. This leads to the following relation between frequency f, propagation speed v, and wavelength λ:

$$v = f\lambda.$$

Example 2.3 According to IEEE 802.11b, Wi-Fi uses a transmission frequency of roughly 2.4 GHz. Assuming a transmission speed of $v = c_0 \approx 300,000$ km/s, we find

$$\lambda = \frac{c_0}{f} \approx \frac{0.3 \cdot 10^9 \, \text{m/s}}{2.4 \cdot 10^9 \, \text{1/s}} = \frac{1}{8} \, \text{m} = 0.125 \, \text{m}.$$

Hence, the wavelength of Wi-Fi is approximately 12.5 cm.

When using an antenna, all electromagnetic waves at the position of the antenna have influence on the voltage level used by the receiver. However, the individual elementary waves, which lead to this observation cannot be extracted by the antenna. That is, the individual frequencies of elementary waves are not accessible. However, there is a strong theorem in math which states that for any periodic and continuous function, a representation as a sum of elementary waves can be defined and calculated. In this case, the frequencies have a special structure dependent on the period length T of the signal. This should be kept in mind, as in many practical cases, this theorem is applied to periodic signals defined from aperiodic signals as the periodic extension of some short interval called window.

Theorem 2.2 *Any periodic, continuous function s of period T can be written as a sum of sines and cosines:*

$$s(t) = \frac{A_0}{2} + \sum_{k=1}^{\infty} (A_k \cos(2\pi k f_0 t) + B_k \sin(2\pi k f_0 t)). \tag{2.2}$$

The frequencies of the sines and cosines are all multiples of a special frequency f_0, called fundamental frequency. *This fundamental frequency is the reciprocal value of the period duration T:*

$$f_0 = \frac{1}{T}.$$

The constant parameters can be given by the following formulas:

$$A_k = \frac{2}{T} \int_0^T s(t) \cos(2\pi k f_0 t) dt$$

$$B_k = \frac{2}{T} \int_0^T s(t) \sin(2\pi k f_0 t) dt.$$

The following example illustrates the use of this theorem in a concrete case:

Example 2.4 Consider the digital signal given by the periodic continuation of

$$s(t) = \text{sgn}(t) \text{ for } t \in (-\pi, \pi), \tag{2.3}$$

where $\text{sgn}(t) = 1$ for $t > 1$, $\text{sgn}(t) = -1$ for $t < 1$, and $\text{sgn}(0) = 0$. Figure 2.5 shows part of this function.

Fig. 2.5 A periodic digital
signal

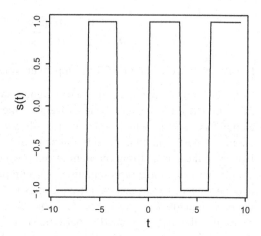

The first task is now to calculate the constants in Eq. (2.2). For this, we need
some preparations given as two lemmata. The first of them states that any integral
of a periodic function over a length of the period duration T is equal. This can be
motivated by the fact that moving the integration bounds in any direction introduces
the same values at one end, which are left out at the opposite end of the interval.

Lemma 2.1 *Let f be periodic with period T. Then*

$$\int_a^{a+T} f(t)dt$$

does not depend on a.

Proof Consider

$$g(a) = \int_a^{a+T} f(t)dt = \int_0^{a+T} f(t)dt - \int_0^a f(t)dt.$$

Let $F(t)$ be a function, such that $F'(t) = f(t)$. Then

$$\frac{d}{da}g(a) = \frac{d}{da}\left((F(a+T) - F(0)) - (F(a) - F(0))\right) = f(a+T) - f(a) = 0.$$

Hence, g is constant in a.

Another fact needed in order to complete the example is given by observing that
integrals of functions over intervals centered at zero can be simplified for even and
odd functions as follows:

Lemma 2.2

$$\int_{-a}^{a} f(t)dt = \begin{cases} 2\int_0^a f(t)dt & for f\ even \\ 0 & for f\ odd. \end{cases}$$

Proof

$$\int_{-a}^{a} f(t)dt = \int_0^a f(t)dt + \int_{-a}^0 f(t)dt \qquad (2.4)$$

$$= \int_0^a f(t)dt + \int_0^a f(-t)dt. \qquad (2.5)$$

For f even, that is, $f(t) = f(-t)$, both summands are equal, and for f odd, that is, $f(-t) = -f(t)$, both summands are equal with different signs and cancel.

With these preparations, it is possible to calculate the Fourier series of $s(t)$. As the function $s(t)$ is odd, that is, $s(-x) = -s(x)$, one can easily calculate that $s(t)\cos(\omega t)$ is odd and $s(t)\sin(\omega t)$ is even for some constant ω. Hence, $A_k = 0$ for all i. For B_i, one calculates with $f_0 = 1/2\pi$

$$B_k = \frac{2}{\pi} \int_0^{\pi} 1 \cdot \sin(2\pi k f_0 t)dt$$

$$= \frac{2}{\pi} \int_0^{\pi} \sin(kt)dt$$

$$= \begin{cases} \frac{4}{k\pi} & for\ k \in \{1, 3, 5, \ldots\} \\ 0 & else. \end{cases}$$

Hence, we can conclude

$$s(t) = \frac{4}{\pi}\left(\sin(t) + \frac{1}{3}\sin 3t + \frac{1}{5}\sin 5t \ldots\right)$$

and give names to partial sums as follows:

$$s_k(t) = \frac{4}{\pi} \sum_{i=0}^{\lfloor k/2 \rfloor} \frac{1}{2k+1}\sin((2k+1)t).$$

Figure 2.6 shows how different partial sums approximate the original signal.

The Fourier series representation of a time-domain signal is very important for wireless communication systems, as we will see later. This representation allows us also to define some very important signal properties.

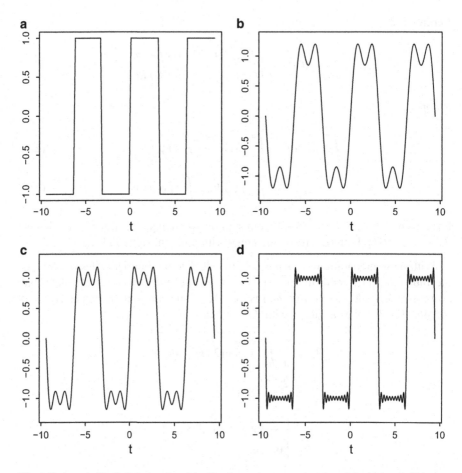

Fig. 2.6 A periodic digital signal and its Fourier series approximation. (**a**) $s(t)$, bandwidth ∞. (**b**) $s_3(t)$, bandwidth $\frac{1}{\pi}$. (**c**) $s_5(t)$, bandwidth $\frac{2}{\pi}$. (**d**) $s_{20}(t)$, bandwidth $\frac{9}{\pi}$

The spectrum of a signal is the set of all frequencies it contains. That is, for the above example, the spectrum is

$$\mathscr{S}(s) = \left\{ \frac{1}{2\pi}, \frac{3}{2\pi}, \frac{5}{2\pi}, \dots \right\} .$$

The *absolute bandwidth* of a signal is defined to be the difference of the largest and smallest element of the spectrum. Hence, for the above example, the absolute bandwidth is infinite. For another very simple signal $s(t) = \sin(2\pi f_0 t)$, the bandwidth is zero, as the spectrum is the one-element set containing f_0.

As most (theoretical) signals have infinite absolute bandwidth, another definition is of high importance. That is 50% bandwidth, which is the length of the part of the spectrum in which 50% of the signal energy is concentrated.

2.2.2 Channel Capacity, Bandwidth, and Data Rate

Closely related to the bandwidth of a signal is the maximal data rate of a channel. Common communication systems are always limited in bandwidth, and this limitation poses an upper bound on the achievable data rate of a communication system. The *data rate* is defined to be the amount of information per time, typically given in bit per second. The maximum data rate that can be achieved on a channel is called *channel capacity*.

Channel capacity, bandwidth, and data rate also depend on the amount of noise energy in comparison to the amount of signal energy. In order to formulate these relations, we first need a suitable way of comparing signal strengths which is given by decibel.

2.2.2.1 Decibels

In wireless transmission systems, signal strength often falls off exponentially with distance. Moreover, the choice of output power is often variable, and hence, the system efficiency is to be given in terms of a relative measure. To account for both aspects, the unit of decibel has been widely known and used.

Definition 2.1 *Decibel* is a relative measure of power on a logarithmic scale using base 10. It expresses relative differences called gain and loss. Therefore, two power variables P_{in} and P_{out} are set into relation with each other as given in

$$G_{dB} = 10 \log_{10} \frac{P_{out}}{P_{in}}.$$

The names P_{in} and P_{out} are chosen, because the most frequent use of decibel is the characterization of gain and loss in amplifiers, where the power levels correspond to input and output of such an amplifier. Due to the choice of a logarithmic unit, cascading of amplifiers leads to simple addition of gain. Hence, two identical amplifiers lead to twice the decibel gain.

In many cases, it is also interesting to have absolute measures of signal energy. Therefore, it is common practice to fix a specific reference energy to be 0 on the invented scale. Most common values for the reference energy are 1 W and 1 mW leading to the units decibel-Watt (dBW) and decibel-milliWatt (dBm).

$$G_{\text{dBW}} = 10 \log_{10} \frac{P}{1\,\text{W}}$$

$$G_{\text{dBm}} = 10 \log_{10} \frac{P}{1\,\text{mW}}.$$

A subtlety with these definitions is often overlooked: if one compares two values on an absolute decibel scale by, for example, subtracting them from each other, then the resulting unit is relative and not absolute. Concretely, let us calculate a dBm $-b$ dBm:

From the definitions, we know that

$$a = 10 \log_{10} \frac{A}{1\,\text{mW}}, \text{ and}$$

$$b = 10 \log_{10} \frac{B}{1\,\text{mW}}.$$

Calculating

$$a - b = 10 \log_{10} \frac{A}{1\,\text{mW}} - 10 \log_{10} \frac{B}{1\,\text{mW}}$$

$$= 10 \left(\log_{10} \frac{A}{1\,\text{mW}} - \log_{10} \frac{B}{1\,\text{mW}} \right)$$

$$= 10 \log_{10} \frac{A}{B}$$

we see that the absolute unit cancels out. Therefore, the unit of $a - b$ is not dBm; it is dB.

Example 2.5 Assume a complex communication setup, where a transmitter is emitting a signal with 1 W signal strength. The first transmission path has a loss of 10 dB; then an amplifier is installed, which has a gain of 15 dB, and then the signal is again emitted over a transmission path with 20 dB loss. The overall transmission loss is then given by

$$-10\,\text{dB} + 15\,\text{dB} - 20\,\text{dB} = -15\,\text{dB},$$

and the absolute output power level P comes out to be

$$G_{\text{dB}} = -15 = 10 \log_{10} \frac{P}{1\,\text{W}}$$

$$P = 10^{-\frac{15}{10}}\,\text{W} \approx 0.031\,\text{W}.$$

Example 2.6 For a real-world application, use a Linux computer with Wi-Fi interface (we assume it to be named wlan0) and issue (as superuser) `iwlist wlan0 scanning`. The adapter returns all visible Wi-Fi access points in the surroundings and especially gives the signal strength in dBm. The output for a single cell looks like:

```
Cell 19 - Address: 00:03:52:AB:88:00
          Channel:1
          Frequency:2.412 GHz (Channel 1)
          Quality=25/70  Signal level=-85 dBm
          Encryption key:off
          ESSID:"lrz"
          Bit Rates:6 Mb/s; 9 Mb/s; 12 Mb/s; 18 Mb/s
          Mode:Master
          [...]
```

To find the absolute signal strength in mW in the signal of this cell, we calculate

$$G_{\text{dBm}} = -85 = 10 \log_{10} \frac{P}{1\,\text{mW}}$$

$$P = 10^{-\frac{85}{10}}\,\text{mW} \approx 3.16 \cdot 10^{-9}\,\text{mW} = 3.16 \cdot 10^{-12}\,\text{W} = 3.16\,\text{pW}.$$

2.2.2.2 Sampling Theorem

For the understanding of the meaning of bandwidth, the following well-known "Sampling Theorem" is very illustrative:

Theorem 2.3 *If a real-valued function $s(t)$ contains no frequencies higher than B, it is completely determined by giving its values at a series of points spaced $\frac{1}{2B}$ apart.*

Proof A very simple proof is given in [10].

In fact, this theorem shows that a bandwidth-limited function has only limited information in the time domain. In other words, a higher sampling frequency than $2B$ does not yield additional information about the signal. So for a given communication system of limited bandwidth B, a uniform discretization with intervals $\frac{1}{2B}$ does suffice to reconstruct the complete signal. This fact is closely related to the following Nyquist theorem.

2.2.2.3 Nyquist Theorem

For a noiseless channel, a famous result of Nyquist states that the maximal signal rate for a channel with bandwidth B is $2B$. Conversely, for a given signal rate of $2B$, a channel with bandwidth B is sufficient to transport the information. Note that the signal rate is not the data rate, as a signal can have more than two states. Taking into account the number of different signal states M, the capacity can be calculated by the following formula:

$$C = 2B \log_2(M).$$

Example 2.7 The telephony system is optimized for human speech and the plain old telephony service (POTS) provides full-duplex transmission of signals ranging from 300 to 3,400 Hz; hence, the POTS provides a bandwidth of 3,100 Hz. Consequently, the capacity of the channel using $M = 8$ different signal states, a common value for modern modems, is

$$C = 2 \cdot 3{,}100\,\text{Hz} \cdot \log_2 8\,\text{bit} = 2 \cdot 3{,}100\,\text{Hz} \cdot 3\,\text{bit} = 18{,}600\,\text{bps}.$$

This number is well known from those days, where the telephony line was (directly) used to get Internet access.

While the Nyquist theorem provides a yardstick for measuring the performance of any communication system in relation to the theoretic maximum, a more specific capacity bound is known for systems following the Shannon communication model with a noise source. The central element in this famous bound is given by comparing the amount of noise energy and the amount of signal energy.

2.2.2.4 Shannon Theorem

For a channel with noise, a result of Shannon gives a formula to calculate the capacity of the channel. Therefore, the amount of noise has to be abstracted into a value called signal-to-noise ratio (SNR). The SNR is basically the quotient of the signal power divided by the noise power, but it is often given in decibel

$$\text{SNR}_{\text{dB}} = 10 \log_{10} \frac{P_{\text{Signal}}}{P_{\text{Noise}}},$$

such that a high SNR means a high-quality channel, because the influence of noise is less. According to Shannon [10], the following formula gives an upper bound on the capacity of a channel with white Gaussian noise:

$$C = B \log_2(1 + \text{SNR})$$

Note that in this formula, the SNR is the quotient and *not* given in decibels.

To illustrate the SNR and its implications on performance, one can assign SNR values to the number of bars on the Windows Wi-Fi indicator. Table 2.3 shows the status of a Wi-Fi connection in specific SNR regions.

Table 2.3 Microsoft
Windows signal strength
symbol and SNR in wireless
LAN 802.11b/g site survey[2]

SNR_{dB}	Wi-Fi indicator bars
>40 dB	5 bars
25–40 dB	3–4 bars
15–25 dB	2 bars
10–15 dB	1 bars
5–10 dB	0 bars

Example 2.8 Assume a Wi-Fi scenario with a noise level of -90 dBm, a received
signal level of -64 dBm, and a channel bandwidth of 22 MHz, as defined in 802.11b.
Then the following calculations give an upper bound on the transmission rate of the
communication channel:

$$SNR_{dB} = -64\,dBm - (-90\,dBm) = 26\,dB$$

$$SNR = 10^{\frac{26}{10}} \approx 398.11$$

$$C = 22\,MHz\,\log_2(1 + 398.11) \approx 175\,Mbps.$$

When the Microsoft Windows signal strength symbol shows one bar, such that
the SNR is between 10 and 15 dB, the channel capacity will be between c and C as
calculated:

$$SNR_c = 10^{\frac{10}{10}} = 10$$

$$c = 22\,MHz\,\log_2(1 + 10) \approx 76.1\,Mbps$$

$$SNR_C = 10^{\frac{15}{10}} \approx 31.62$$

$$C = 22\,MHz\,\log_2(1 + 31.62) \approx 110.6\,Mbps.$$

Of course, these upper bounds will never be reached by a real system, because
there are many other sources of signal degradation, and also the transmitter and
receiver will never be lossless. Still these values provide a convenient yardstick for
the efficiency of a transmission system.

2.2.3   Antennas and Signal Propagation

Antennas are the part of every electromagnetic transmission system, which bridges
between the medium (e.g., air) and electrical signals. By an antenna, a specific
type of electrical signal can be transformed to an electromagnetic wave, which will

propagate from the antenna through the medium. Therefore, the electrical signal has to be an electromagnetic signal of the type as above (2.1). The properties of antennas imply that only signals with high enough frequencies can be emitted. In general, an antenna is symmetric in the sense that the properties with respect to detecting, collecting, and receiving signals from electromagnetic waves are equal to the properties of the antenna with respect to emitting waves. This result follows from the principle of reciprocity in electromagnetism. Reciprocity in this case means that the location of an experiment's emitter and detector can be exchanged without changing the result. That is, given an oscillating current at some point p and a measurement of the electrical field at some point q, moving the oscillating current to location q and measuring at location p give the same result.

Definition 2.2 An *antenna* is an electrical conductor used either for radiation electromagnetic energy or for collecting radiated electromagnetic energy [11, Chap. 5.1].

A special type of antenna is the *isotropic antenna*, which is an ideal antenna without geometric extent (e.g., a point). Though isotropic antennas do not exist, they have plenty of applications in modeling and comparison of real-world antennas. The radiation of an isotropic antenna is evenly spread over the surface of a sphere. Signals detected with an isotropic antenna consist of the complete radio energy located at a point in space over time.

Real antennas always have some geometric extent and often have different characteristics with respect to different radiation directions. These characteristics are often given as the deformation of a unit sphere, most often given as a two-dimensional projection. Figure 2.7 gives an example for the Hertzian dipole. The dipole antenna is one of the simplest constructions and consists of two identical pieces of metal. The current of the transmitter is applied between the two pieces of the antenna, and the receiver detects electromagnetic waves by observing the voltage between these two parts.

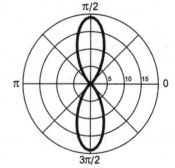

Fig. 2.7 A radiation pattern
for an optimal dipole antenna
for the case of Wi-Fi

A more specialized measure of antenna directivity is given by the *beam width* of an antenna, which is defined to be the length of the maximal interval of angles, where the antenna is emitting more than half of the energy it is emitting into the most preferred direction. This is sometimes referred to as the half-power beam width to clarify the factor of $1/2$.

Example 2.9 Assume a perfect dipole antenna in upright position of length $2L$, where an electrical power of I_m is flowing in the vertical direction. Then the average radiated power density can be given by the formula

$$P_d = \frac{15I_m^2}{\pi r^2}\left(\frac{\cos(\beta L\Theta) - \cos(\beta L)}{\sin(\Theta)}\right)^2,$$

where β is a shortcut for $\beta = 2\pi/\lambda$. Setting $r = 1$, $I_m = 1$, $f = 2.4\,\text{GHz}$, $L = \lambda/2$, $\Theta \in [0, 2\pi]$ gives a radiation pattern as depicted in Fig. 2.7. This example has a beam width of $46.8°$.

This measure characterizes the geometric characteristics of the antenna quite well. The geometric shape in Fig. 2.7 is typical for this type of antenna, and the beam width is a good single-value characterization of the directionality of a dipole antenna.

Another measure, called *antenna gain*, characterizes the efficiency of transmission into the different directions. Antenna gain is defined to be the quotient of the power output into a specific direction as compared to the power output of an idealized isotropic antenna. If no direction is mentioned, antenna gain typically refers to the maximal antenna gain.

The antenna gain is again closely related with another geometric measure, called the *effective area* or *aperture* of the antenna. The aperture is the area (perpendicular to the propagation direction), which "contains" the amount of radiation energy, which is converted into current by the antenna. So the effective area is always smaller than the physical area of the antenna leading to another relative measure: *aperture efficiency*, which is defined to be the effective area divided by the physical area:

$$e = \frac{A_{\text{eff}}}{A_{\text{phys}}}.$$

This measure is always between 0 and 1, typical values for real-world antennas range from 0.45 to 0.7.

Of course, the antenna aperture is closely related to the antenna gain by the following formula:

$$G = \frac{4\pi A_{\text{eff}}}{\lambda^2}. \tag{2.6}$$

2.2.3.1 Propagation

Propagation is a complex topic, which is fundamental to the design and understanding of mobile communication systems. Propagation of radio waves is the set of physical laws, which defines how the radio wave actually travels through media and especially how the radio wave is affected when the transmission medium changes.

Basically, three modes of propagation have to be distinguished:

- Ground wave propagation, where the electromagnetic wave merely follows the shape of the world.
- Sky wave propagation, where the electromagnetic wave is reflected by the ionosphere and sometimes kept in a corridor such that it reaches its destination after some reflections.
- Line-of-sight (LOS) propagation, where the electromagnetic wave does not change its direction while changing medium.

The *mode of propagation* mainly depends on the frequency of the waves. As a rule of thumb, ground wave propagation is applicable for frequencies below 2 MHz, sky wave propagation is applicable for frequencies between 2 and 30 MHz, and LOS propagation is applicable for frequencies over 30 MHz [11, Chap. 5.2].

For *ground wave propagation*, the wave follows the rounded shape of the earth due to refraction effects (see Fig. 2.8a). The refractive index of the atmosphere continuously decreases with height, such that radio waves travel slower near the earth leading to a slight bend towards the earth.

As these effects lead to higher viewing distances as compared to a direct line on the surface of the earth, the following formula gives an approximate maximal radio propagation distance using an antenna of height h:

$$d = 3.57\sqrt{Kh}.$$

For light, $K = 1$, for radio $\frac{4}{3}$ is a common value. However, the refractive index of a medium depends on the frequency, and hence, the above expression can only be seen as a coarse approximation. To find out the maximum distance between two

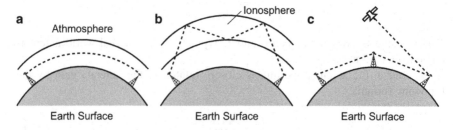

Fig. 2.8 Different modes of propagation for radio transmission. (**a**) Ground wave propagation. (**b**) Sky wave propagation. (**c**) Line-of-sight propagation

antennas of given heights h_1 and h_2, it is easy to see that they can communicate if and only if the circles given by the maximal radio propagation distances of both antennas intersect. Because if they are farther away, the rays would have to cross the earth, and if they are nearer, they can easily travel through the air. The maximal radio communication distance between two antennas of known heights is thus given by

$$d_{1,2} = 3.57(\sqrt{Kh_1} + \sqrt{Kh_2}).$$

In case of *sky wave propagation*, the waves emitted by an earth-based antenna are reflected by the ionosphere and the earth surface and are thus able to travel very large distances. Figure 2.8b depicts an example.

The most complex and for mobile communication systems most important propagation mode is LOS propagation. The high-frequency radio waves are not reflected by the ionosphere and also do not obtain enough bending to follow the earth curvature. Hence, communication must be LOS. Figure 2.8c shows two basic examples of LOS propagation, which occurs between base stations and satellites as well as between two base stations, when the transmission frequencies are high.

2.2.3.2 Path Loss

In wireless communication systems, the energy of a signal, which is sent out at a specific place using an idealized isotropic antenna, has to keep constant. Hence, this constant amount of energy is spread out over a sphere, whose radius is growing with time based on the propagation speed of the electromagnetic wave inside the medium as depicted in Fig. 2.9. This picture also marks small fractions of the signal sphere with aperture. This is intended to visualize how the amount of energy decreases for a receiving antenna of constant aperture. Without other impairments, the free space loss of an isotropic antenna is given by

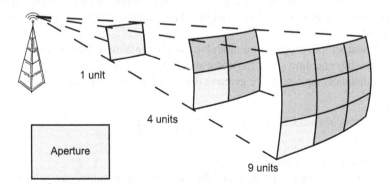

Fig. 2.9 The inverse square law and the amount of energy a constant aperture antenna can receive

$$\frac{P_t}{P_r} = \frac{(4\pi d)^2}{\lambda^2}.$$

In this equation, the left side is the quotient of the transmitted power P_t and the received power P_r; the right side contains the propagation distance d and the wavelength λ. This equation is also often given on a decibel scale, where it reads

$$L_{dB} = 10\log_{10}\left(\frac{P_t}{P_r}\right) = 20\log_{10}\left(\frac{4\pi d}{\lambda}\right) \tag{2.7}$$

$$= -20\log_{10}\lambda + 20\log_{10}d + 21.98\,\text{dB}. \tag{2.8}$$

For non-isotropic antennas, one can use either the antenna gain or aperture, which is linked using Eq. (2.6):

$$\frac{P_t}{P_r} = \frac{(4\pi d)^2}{G_r G_t \lambda^2} = \frac{(\lambda d)^2}{A_r A_t}.$$

This family of equations is known as Friis transmission equations, and they are valid, if the following conditions are met:

- The transmission distance exceeds the wavelength by magnitudes, i.e., $d \gg \lambda$,
- the antennas are correctly aligned and polarized, and
- the transmission bandwidth is narrow enough to allow for one wavelength being used.

2.2.3.3 Statistical Signal Propagation Models

To cover the very complex situation of transmission and channel loss in multipath scenarios inherent to indoor location-based services, there are plenty of propagation models, which subsume all unknown effects into values, which can empirically be estimated.

The most simple propagation model is the propagation in free space (vacuum). The following equation gives the corresponding equation (λ is the wavelength, d denotes distance, the results are given in dB) modeling the physical relationships:

$$E(d) = 10\log_{10}\left(\frac{4\pi d}{\lambda}\right)^2.$$

This essentially is a recast of Eq. (2.7).

This equation consists of an application of the inverse square law which states that radiated energy of a point source fades with the square of the distance, as the energy is distributed across the surface of a sphere ($4\pi r^2$), which grows quadratically with the radius. The other component is modeling the aperture of the receiving antenna and is frequency dependent.

For narrowband signals, it is customary to simplify the discussion to signals of a single constant frequency $f = \frac{c}{\lambda}$. This introduces some inaccuracies, but they are often smaller than all other sources of perturbations of non-modeled system properties including antenna gain, propagation path, reflection, distortion, etc.

As a first step, all these unknown sources of signal perturbation are collected into a single value, the propagation exponent (^{2} in the free space equation above). Measuring a reference value $E(d_0)$ at a fixed reference distance d_0, the so-called one-slope model, is derived in which the two model parameters α and $E(d_0)$ are derived from measurements

$$E(d,\alpha) = E(d_0) - 10\alpha \log_{10}\left(\frac{d}{d_0}\right).$$

The reference distance is often taken as $1\,m$. Typical values for α are given in the literature [7] ranging from 1.8 to 3. The influence of specific material on Wi-Fi propagation is, for example, studied in [14].

The next obvious refinement of this model is to add support by a floorplan. In a floorplan, we could calculate the number of walls which have to be penetrated and calculate a common wall attenuation factor. Taking this refinement, we end up with the Motley–Keenan model [4], which is used, for example, in RADAR [1]. The refined model equation is given below, where the constant wall attenuation factor is given as W and n denotes the number of walls between receiver and transmitter:

$$E(d,\alpha,n) = E(d_0) - 10\alpha \log_{10}\left(\frac{d}{d_0}\right) - nW.$$

An illustrative example of the Motley–Keenan propagation model is depicted in Fig. 2.10. The depicted signal strength degrades using the free space model and additionally drops at the location of walls.

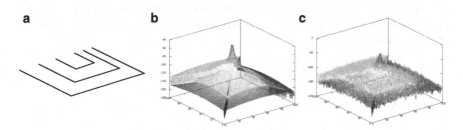

Fig. 2.10 Motley–Keenan propagation model for a simple floorplan. (**a**) Floorplan. (**b**) Motley–Keenan model without noise. (**c**) Motley–Keenan model with white Gaussian noise

Based on the previous propagation models, which are purely based on the direct connecting line between sender and receiver, several models have been proposed and discussed, which take into account other propagation paths. The problem with simulating the physical properties are twofold: first of all, the available floorplans are not sufficiently correct, and on the other hand, the solvation of the involved differential equations with finite difference methods or finite element methods is computationally very expensive. A good way in the middle is the adoption of raytracing techniques, as they give good results with medium calculational overhead [12]. Another technique, which further reduces from the complexities of raytracing is the calculation of so-called dominant paths [13]. A dominant path in a wireless transmission scenario is the path of smallest fading and therefore the path on which the strongest signal in multipath scenarios travels between the transmitter and the receiver. The length of this dominant path is then used as the distance parameter for some distance-based propagation model.

2.2.4   Modulation

Modulation is the process of changing the parameters of a signal in order to transmit information. For signals, which are based on waves, the three parameters of an elementary wave, namely, amplitude, phase, and frequency, can be changed according to the information to be transmitted. This results in three basic modulation schemes in which the signal information is transmitted by adapting one of these three parameters. Further, it is possible to combine several such schemes to produce higher-order schemes in which several parameters of the wave are changed at the same time in order to transmit more information at a given time.

2.2.4.1   Amplitude Modulation

The main idea of *amplitude modulation (AM)* is to encode information into time-domain changes of the amplitude of a signal. In amplitude modulation, typically two signals are intermixed: first of all, a carrier wave signal $s_c(t)$ is generated, which often is a simple sinus wave:

$$s_c(t) = A_c \sin(2\pi f_c t).$$

The modulation equation

$$p(t) = (A_c + s(t)) \sin(2\pi f_c t) \qquad (2.9)$$

intermixes this carrier signal with the information signal $s(t)$ in a way, such that the resulting signal can be decoded and the information signal $s(t)$ can (approximately) be reconstructed. Note that in the literature and in drawings, both waves often have some synchronization property, such that complete wave cycles of the information signal are encoded onto multiple complete wave cycles of the carrier signal. This is in fact not needed and would be extremely difficult to realize; however, we will follow this common simplification in all illustrations (Figs. 2.11 and 2.12).

Amplitude modulation is typically adding sidebands next to the carrier frequency f_c. In other words, the spectrum of the resulting signal after AM modulation contains the carrier frequency as well as other frequencies below and above the carrier frequency, which solely depend on the data signal used during modulation. This effect is best derived by calculating the modulation spectrum in an accessible example:

Example 2.10 Assume an orchestra installs a tuning sender, which uses amplitude modulation to distribute the chamber tone of $f_s = 440\,\mathrm{Hz}$. The signal has the form

$$s(t) = A_s \sin(2\pi f_s t).$$

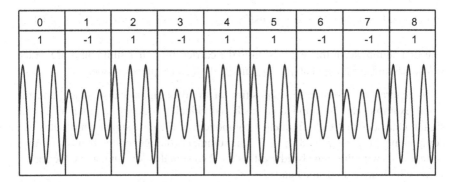

0	1	2	3	4	5	6	7	8
1	-1	1	-1	1	1	-1	-1	1

Fig. 2.11 A simple AM example for a digital signal

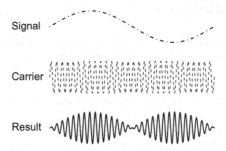

Signal

Carrier

Result

Fig. 2.12 Amplitude modulation of an analog signal

It is now common practice to normalize the modulation equation (2.9), such that $A_c = 1$. This is done by introducing the *modulation index*

$$m = \frac{A_s}{A_c}, 0 \le m \le 1.$$

Using this notation, one calculates

$$\begin{aligned} p(t) &= (A_c + A_s \sin(2\pi f_s t)) \sin(2\pi f_c t) \\ &= A_c (1 + m \sin(2\pi f_s t)) \sin(2\pi f_c t) \\ &= A_c \sin(2\pi f_c t) + A_c m \sin(2\pi f_c t) \sin(2\pi f_s t). \end{aligned}$$

Using now the well-known fact

$$\sin(a) \sin(b) = \frac{1}{2}(\cos(b - a) - \cos(b + a)),$$

one gets the following representation as cosine waves:

$$p(t) = A_c \sin(2\pi f_c t) + A_c \frac{m}{2} \cos(2\pi (f_c - f_s)t) - A_c \frac{m}{2} \cos(2\pi (f_c + f_s)t).$$

The first summand in this expression is the carrier, the other summands are called *sidebands* and are located at $f_c - f_s$ and $f_c + f_s$ in frequency space.

In general, the Fourier decomposition of the data signal can be used, and the given example can be applied to each individual frequency component of the signal. One sees that each frequency component of the data signal contributes two components at $f_c - f_s$ and $f_c + f_s$. In summary, this generates bands of signal components with the same bandwidth as the data signal below and above the carrier wave, which are called side bands.

2.2.4.2 Frequency Modulation

Frequency modulation (FM) is changing the frequency of a carrier signal according to an information signal. As with amplitude modulation, a carrier wave signal $s_c(t)$ is generated, which is a simple sinus wave:

$$s_c(t) = A_c \sin(2\pi f_c t).$$

If we assume that the baseband information signal $s(t)$ is limited (or normalized) to contain only values from the interval $[-1, 1]$, we can give the following modulation equation for frequency modulation:

$$p(t) = A_c \sin\left(2\pi [f_c + f_\Delta s(t)]t\right). \tag{2.10}$$

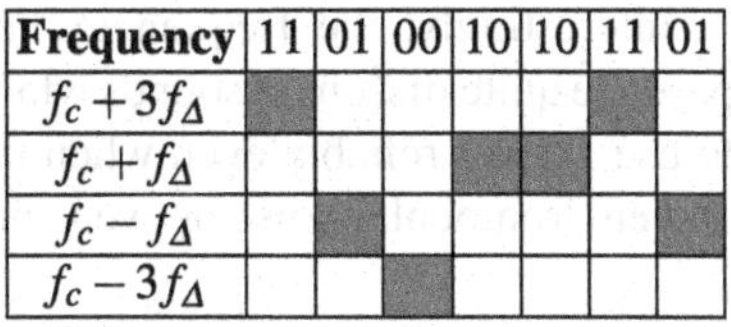

Fig. 2.13 An example of MFSK using four frequencies, encoding two bits at a time

Frequency	11	01	00	10	10	11	01
$f_c + 3f_\Delta$	■					■	
$f_c + f_\Delta$				■	■		
$f_c - f_\Delta$		■					■
$f_c - 3f_\Delta$			■				

Frequency-modulated analog signals have infinite bandwidth, but frequency modulation creates only few strong sidebands above and below the carrier frequency. Carson's rule gives an upper bound to the 90% bandwidth of such a signal to be

$$B = 2(f_m + f_\Delta),$$

where f_m is the maximum modulation frequency, that is, the maximal frequency of the baseband signal that gets modulated onto the carrier wave. This means that 90% of the energy of a FM sender are contained in the interval $[f_m - f_\Delta, f_m + f_\Delta]$.

The value f_Δ gives the maximal deviation of the signal from the carrier frequency f_c. If $s(t)$ is a binary function taking two values -1 and 1, we can write the equation of frequency modulation as follows:

$$s(t) = A_c \sin\left(2\pi\left(f_c + \frac{1}{2}(d(t) + 1)f_\Delta\right)t\right)$$

In this case, frequency modulation is also known as *Binary Frequency Shift Keying (BFSK)*.

There is a straightforward extension to BFSK known as *Multiple Frequency Shift Keying (MFSK)*, where more than two frequencies are used to transmit more than one bit at a time. In this case, the signal is composed of a set of signal elements at those various frequencies, where a constant-frequency tone at a given frequency encodes more than one bit. An example using four various frequencies, encoding two bits at a time, hence having half the symbol rate of the bitstream, is given in Fig. 2.13.

Example 2.11 A nice example of frequency modulation using multiple frequencies is *dual-tone multi-frequency (DTMF)*, a system for communication numbers through overlays of two sinusoids in telephony applications. Therefore, two fixed-frequency sinus signals are generated, based on the assignment matrix in Table 2.4, and these two signals are then added up. This system communicates four bits at a time.

Table 2.4 DTMF frequency assignment. Touching a key on a DTMF phone sends the sum of two sinusoids of the given frequencies through the line

	1,209 Hz	1,336 Hz	1,477 Hz	1,633 Hz
697 Hz	1	2	3	A
770 Hz	4	5	6	B
852 Hz	7	8	9	C
941 Hz	*	0	#	D

Interestingly, the frequencies in Table 2.4 have been chosen such that they produce quite dissonant sounds which rarely occur in nature. Therefore, it is possible to use DTMF reliably even when there is a microphone attached to the phone line and environmental noise or even spoken language is transmitted on the line.

2.2.4.3 Phase Modulation: Phase Shift Keying

Table 2.5 Identifying true with 1 and false with -1 makes XOR negated multiplication

Boolean			$\{-1, 1\}$		
A	B	A XOR B	A	B	$-AB$
False	False	False	-1	-1	-1
False	True	True	-1	1	1
True	False	True	1	-1	1
True	True	False	1	1	-1

Phase Modulation is one of the most important types of modulation for binary signals. This section only covers digital signals, and hence, the term phase shift keying (PSK) is preferred. In PSK of binary signals, a bitstream is encoded into changes of the phase of a signal. Throughout this chapter, we identify bits with the set $\{-1, 1\}$, such that exclusive or (XOR) becomes multiplication (see Table 2.5).

In a simple version of PSK, called Binary Phase Shift Keying (BPSK), two different phase angles are assigned with the two states from the set $\{-1, 1\}$. For two such states, it is easy to write BPSK as in the following formula:

$$s(t) = Ad(t) \cos(2\pi f_0 t).$$

Here, the phase difference between a -1 and a 1 is exactly π leading to a sign inversion of the cosinus function. As a signal in time, this leads to phase jumps as depicted in Fig. 2.14a.

There is a variant of BPSK, which is not encoding the data directly but rather changes in the data. As long as the data sequence does not change value, the phase does also not change. When the data changes from -1 to 1 or vice versa, then the phase of the BPSK signal is simply inverted. This scheme is known as *Differential Binary Phase Shift Keying (DBPSK)*.

BPSK is only transmitting one bit per symbol as it is only using a phase change of π. A slightly more complex encoding technique is known as *Quadrature Phase Shift Keying (QPSK)*. QPSK uses four phases with a spacing of $\pi/2$ transmitting two bits per symbol. As a prerequisite to this type of encoding, we need some simple knowledge of complex numbers, which can be found in any basic textbook on analysis.

We will often use the polar representation

$$z = e^{i\phi}$$

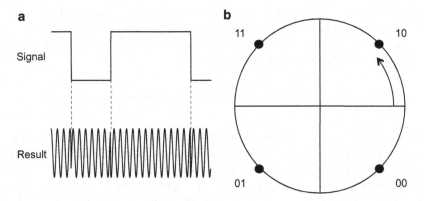

Fig. 2.14 Different PSK variants. (**a**) BPSK-encoded signal. (**b**) QPSK encoding scheme

as well as the Cartesian representation

$$z = a + ib = \sqrt{a^2 + b^2}(\cos(\phi) + i \sin(\phi)).$$

Basically, a QPSK-encoded signal encodes two bits at once. A sequence of two bits is often called dibit. The QPSK encoding can be given by the following set of formulas

$$s(t) = \begin{cases} A\cos(2\pi f_0 t + \frac{\pi}{4}), \text{ if the dibit is } 11 \\ A\cos(2\pi f_0 t + \frac{3\pi}{4}), \text{ if the dibit is } 01 \\ A\cos(2\pi f_0 t - \frac{3\pi}{4}), \text{ if the dibit is } 00 \\ A\cos(2\pi f_0 t - \frac{\pi}{4}), \text{ if the dibit is } 10. \end{cases} \tag{2.11}$$

However, it is much easier to understand if the phase is drawn as a complex number in a polar coordinate system as in Fig. 2.14b. The carrier is rotating around the center, and the location of the phase changes is depicted and located at the four diagonals splitting each quadrant. For a given transmission dibit, the signal takes a complete round in the circle starting and ending at the phase indicated by the dibit. This representation also sheds light on the choice of phase-dibit pairs in Formula (2.11). The first bit of the dibit specifies the "X-axis" and the second one the "Y-axis" of the phase change location in complex coordinates.

For QPSK encoding, the bit stream is usually split into two streams called I-stream (in-phase) and Q (quadrature)-stream. This is done by a 2-bit serial-to-parallel converter. This splitting of both streams results in half bitrates for the I-stream and Q-stream. A single oscillator is generating the carrier signal, and this carrier signal is multiplied with both streams in the Q-stream after applying a phase inversion. Both signals are then combined by addition and transmitted. From this perspective, it becomes clear that QPSK is the same as applying BPSK on the individual split streams using the same, but phase-inversed, carrier.

Denoting the in-phase bitstream by $I(t)$ and the quadrature bitstream by $Q(t)$, we can write QPSK modulation as follows:

$$s(t) = \frac{1}{\sqrt{2}} I(t) \cos(2\pi f_0 t) - \frac{1}{\sqrt{2}} Q(t) \sin(2\pi f_0 t).$$

A simple extension to QPSK limits phase jumps to $\pi/2$ and hence facilitates actual implementation for high data rates. Therefore, after splitting the bitstream into $I(t)$ and $Q(t)$, the quadrature stream is delayed by one bit duration of the original stream, hence a half bit duration of the split streams. In this way, only one bit can change at any time, and hence, the phase jumps cannot exceed $\pi/2$. This technique is called OQPSK which stands for offset QPSK or orthogonal QPSK. The modulation equation for OQPSK is pretty similar to the one of QPSK given above; let T_b denote a bit duration of the unsplit input signal, then

$$s(t) = \frac{1}{\sqrt{2}} I(t) \cos(2\pi f_0 t) - \frac{1}{\sqrt{2}} Q(t - T_b) \sin(2\pi f_0 t).$$

The most important achievement of OQPSK is that no symbol change is crossing the point 0 in the complex plane. Such a zero crossing would lead to a drop of the signal amplitude and renders synchronization and carrier detection at the receiver much more difficult.

The presented phase modulation schemes can be ordered by the size of the maximal phase jumps occurring in the modulated signal. Due to electronic limitations, it is, of course, a design goal to minimize either the number of phase jumps or the size of phase jumps, as they are difficult to realize in hardware. The presented phase modulation schemes each reduce the size of the phase jump. However, it is possible to have no phase jumps at all. These modulation schemes are called *Continuous Phase Shift Keying (CPSK)*. An example of such a modulation scheme is known as *Minimum Shift Keying (MSK)*, and a variant, namely, Gaussian Minimum Shift Keying (GMSK), has been widely used in the GSM cellular network.

MSK works identical to OQPSK, first splitting the signal into half-rate I and a delayed Q-streams, respectively. However, these streams are then manipulated in a way such that the rectangular pulses in the I- and Q-streams are replaced by half-cycle sinusoids. Therefore, the I-stream is multiplied by $\cos(\pi t/2T_b)$, where T_b denotes the bit duration. The (delayed) Q-stream is multiplied by $\sin(\pi t/2T_b)$. It can now be shown that phase changes are linear and limited to $\pi/2$ over a bit duration T_b. Especially, the phase is not jumping anymore.

The basic idea of MSK is replacing the bad-behaved orthogonal pulse structure with a well-behaved shape such as a sinusoidal one. It is, of course, possible to use other shapes. In practice, this shape is often a Gaussian-filter of the digital pulse signal leading to a modulation scheme called *GMSK*. The discussion of the details of GMSK is beyond the scope of this chapter.

2.2.4.4   Combinations of Modulation Schemes

The modulations schemes presented in the previous sections all center on one aspect of the transmission signal: either the amplitude or the frequency or the phase. It is of course possible to mix up individual modulation schemes. The most prominent type of a combined modulation scheme is *quadrature amplitude modulation (QAM)*. In QAM, more than two bits can be transmitted by constructing a phase-amplitude diagram as depicted in Fig. 2.15. In this situation, transmission symbols are distinguished not only by their individual phase with respect to a common synchronization but also by their amplitude. This scheme can arbitrarily be extended by adding phase angles or amplitude levels. However, the receiver will have to be able to distinguish them clearly. As a result, high-order QAM only appears on cable transmission as, for example, in DVB-C with QAM-256, where each modulation symbol carries eight bits. For mobile communication systems, QAM is limited by signal degradation highly affecting the amplitude of a received signal. Therefore, QAM variants are often adaptively scaled using many signal states for nearby pairs of transmitter and receiver falling back to lower-order QAM deployments for distant transmitters. Notably it is possible to transmit high-order QAM symbol and still decode the low-order sub-signal. Therefore, additional information for nearby stations can be added to the transmission symbol though it won't be decodable for distant stations.

Therefore, the ordering of the symbols has to be made as depicted in Fig. 2.15. A receiver that is able to assess the quadrant of an incoming symbol will be able to decode the first two bits of the four bits actually transmitted. This property is often used for video transmission in which the baseline information is encoded into the lower-order bits of the signal, while the higher-order bits contain additional information increasing the video quality. Figure 2.16 depicts examples of the different modulation schemes discussed above for the same data sequence. One can easily see how QPSK decreased the number and size of jumps in comparison to BPSK and that MSK is continuous, e.g., the signal does not jump between two values.

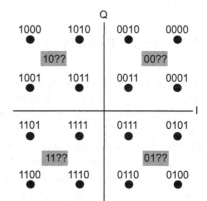

Fig. 2.15 Quadrature-amplitude-modulation QAM16 transmitting four bit per symbol

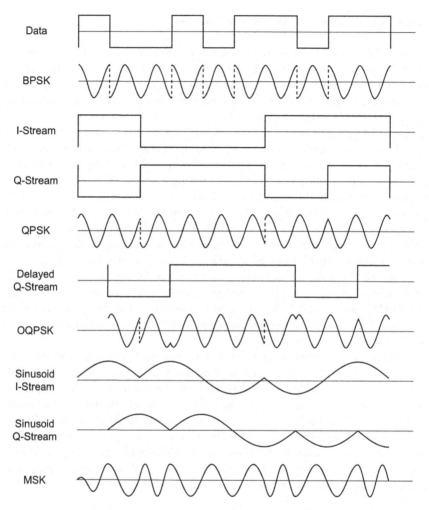

Fig. 2.16 Examples of different modulation schemes for the bit sequence "1, 0, 0, 1, 0, 1, 1, 0, 1, 1"

2.3　Sensor Technology for Positioning

The preceding section has explained the basics of mobile communication systems. These are very central to location-based services, as they are one of the most successful techniques when it comes to inferring the location of a mobile device. Outside, all positioning systems based on satellite technology including GPS, GLONASS, and the upcoming Galileo system are based on an infrastructure of wireless communication endpoints. The satellites send out signals and information, which can be used to infer the location of a device all over the earth. Though wireless communication technologies are among the most successful techniques for determining the position of a mobile device inside a building, other information relevant to the positioning problem has been used successfully. Especially the distribution of smartphones including lots of sensors has led to the ubiquitous possibility to access measurements from other domains. This section brings together the fundamental aspects of measurements possible with current mobile devices in practice.

2.3.1 Time Synchronization and Time Measurement

One of the most basic problems for indoor geolocation is time measurement and time synchronization. A lot of techniques for determining location are based on observing timings of signals with limited propagation speed. The basic approach is often to estimate the distance between a mobile station and known reference stations from the time that a specific signal needs to travel in between. Therefore, some network elements have to share the same time in great accuracy. This section explains the basic algorithms used to create a common sense of time in a distributed system of nodes.

These signals used for distance estimation include electromagnetic waves and light traveling at approximately 300,000 km/s as well as sound with a comparably low propagation speed of roughly 343 m/s. For inferring location information with these signals, we can observe a local time difference between two events such as sending out a signal and receiving the reflection of the very same signal. Therefore, we must be able to measure time between events with challenging accuracy. On the other hand, we can measure the relative time between observing two events such as receiving a message from two different senders. Therefore, however, the senders need to have a correct and consistent time. This leads to the problem of distributed time synchronization.

For computer systems, it is customary to use very simple clock models to model the behavior and errors of two different clocks. These two clocks can be one ideal clock compared to a real clock as well as two erroneous clocks in different devices. The most common clock model for computer systems consists of the following two-parameter representation:

$$C(t) = \theta + f \cdot t.$$

In this expression, $C(t)$ denotes the time as reported by the clock, while the parameter t represents an ideal global time. In other words, this function captures how the clock behaves with respect to time going by reflected by the parameter t being increased linearly. The model parameter θ denotes the *clock offset* that is the time that this clock is different from the reference at $t = 0$. The parameter f models the *clock skew* or *slope*. The slope measures whether this clock is going faster or slower than the optimal clock. This representation is usually used for expressing the comparison of a given real clock with respect to an ideal clock running with offset 0 and slope 1 for which the clock function is the identity, e.g., $C(t) = t$.

When using this clock model relatively to measure the difference between two non-ideal clocks, one can identify both expressions and collect the differences between the two offsets and the two slopes with respect to a perfect clock into a relative value as in

$$C_B(t) = \theta_{AB} + f_{AB} \cdot C_A(t).$$

Given two expressions $C_A(t) = \theta_A + f_A t$ and $C_B(t) = \theta_B + f_B t$, the relative values can be calculated as

$$f_{AB} = \frac{f_B}{f_A} \text{ and}$$

$$\theta_{AB} = \theta_B - f_{AB}\theta_A.$$

Two clocks are in perfect synchronization, of course, when their relative slope is $f_{AB} = 1$, and their relative offset is $\theta_{AB} = 0$.

Figure 2.17 depicts the effect of the model parameters. In the diagram, the X-axis covers the real time (optimal, true time). In this graph, the slope of the clock amounts to the slope of the line and the offset to the value at $t = 0$.

The *distributed clock synchronization* problem for N nodes of a computer network is given by estimating the relative offset and slope of all pairs of nodes such that all nodes can correct their clocks such that all nodes have a common, consistent time. This time need not be correct in a global sense. In contrast to that, the *central clock synchronization* problem of N nodes in a distributed network is the problem of estimating the offset of each node with respect to one timing master node and correcting the local clocks in a manner such that all nodes share the same time with the central clock master. In reality, both approaches come up in a mixed service. Some reference nodes are attached to atomic clocks and provide a very good time estimate. However, even these clocks cannot be in perfect synchronization.

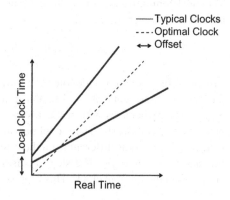

Fig. 2.17 Different clocks

Hence, these clocks use a distributed clock synchronization algorithm to correct their own minor errors. This time is than repeatedly distributed to a more distant set of nodes called strata. Inside each stratum, distributed synchronization is done, and with respect to the higher stratum, a central clock synchronization is performed. Figure 2.18 depicts this organizational structure.

The most important problems for measuring times with computer systems are rooted in the best-effort nature of computer systems. The actual time of execution of a particular piece of software such as an interrupt is not known, and the delays are not constant. They depend on the number of concurring events, the load of the computer system, and the scheduling algorithm used to resolve concurrency. The most important times for measuring the transmission time of a packet transmitted over a wireless networking interface can be detailed as follows:

- *Preparational Time:* The amount of time needed to encode the message into a suitable data structure and hand it over to the networking device, e.g., by activating an interrupt.
- *Medium Access Time:* The waiting time defined on the MAC layer to resolve collisions. This time is highly variable as it depends on the network activities of other nodes and often contains random back-off components.
- *Transmission Time:* The time that the signal travels between the sender and the receiver. This is mainly given by the propagation speed of the signal and the length of the propagation path, which can vary due to multipath effects.
- *Receiving Delay*: The time to decode the transmitted signal, apply error detection and correction, and process it before handing it over to the higher layers.
- *Dispatching Delay*: The time between providing a complete packet inside the network device and reception of this packet at the application layer by an application.

Figure 2.19 depicts these sources of delay for network transmission. The preparation time and dispatching time depend very much on the operating system and computer architecture, while the medium access time and the receiving time

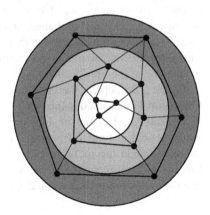

Fig. 2.18 Organizational structure of clock synchronization

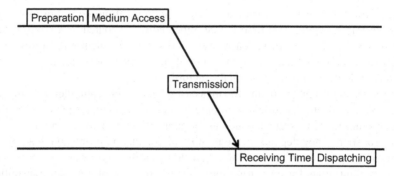

Fig. 2.19 Different sources of delay for time synchronization using wireless communication systems

depend on the implementation of the MAC layer. The access time depends on the multiple access protocol in use and the surroundings, while the reception time is mainly defined by the error detection and error correction scheme.

In real-world applications, it is very difficult to capture these error times and end up with a reasonable time measurement. Therefore, a lot of work is done towards late timestamping of outgoing packets and early timestamping of incoming packets inside network hardware. The hardware driver of the sending system should be able to get a nanosecond estimate of the actual time the first bit was modulated onto the antenna, and, vice versa, the hardware driver of the receiving system should be able to access such a timestamp for the incoming packet together with the decoded packet. In advanced scenarios, the outgoing timestamp should be transmitted already in the same packet and, thus, be accessible to the receiver.

2.3.1.1   Two-Way Message Exchange

For the distributed clock synchronization problem in computer networks, we can assume that all nodes are able to send a message to all other nodes. Following the two-way message exchange principle, this is used to measure the relative parameters under the assumption that the network transmission delay is symmetric. This subsumes into assuming that the time a packet travels from one point to another is the same as the time needed for the way back.

Figure 2.20 depicts the basic situation. Node B sends a packet to node A at timestamp T_1 measured in the clock of B. Node A receives this packet at timestamp T_2 measured in the clock of A. Node A decodes the packet and prepares a response, which is sent at time T_3 measured using the clock of A. This timestamp is also included inside the message and transmitted back to B. B can then record the timestamp T_4 of receiving the answer using its local clock. Under the assumption that the transmission time is equal, the two triangles are of equal shape, and the time

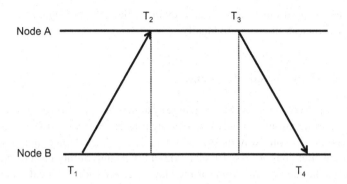

Fig. 2.20 Two-way message exchange mechanism

needed to transmit the packet can be obtained easily from the following formula:

$$\Delta t = \frac{(T4 - T1) - (T3 - T2)}{2}.$$

With this number, node B can correct the local clock by setting it to

$$T_3 + \Delta t$$

at timestamp T_4. Then, node B has the same time as compared to node A. However, this algorithm does not involve possible slopes and is only used for a single measurement. As stated earlier, there are many sources of perturbation for the actual transmission times and time measurements, and therefore, multiple measurements should be used to update time and slope. Therefore, the diagram of Fig. 2.20 can be expressed in two equations involving the relative offset and slope between the two clocks. Assume that k instances of the two-message exchange have been performed for stationary nodes resulting in timestamps $T_{1,k} \ldots T_{4,k}$. Then, we can model the delay as being composed of a fixed part τ equal for both directions and two zero mean random variables X_k and Y_k, each for one direction. This leads to the following equations with unknowns θ, τ, and f:

$$T_{2,k} = f(T_{1,k} + \tau + X_k) + \theta \tag{2.12}$$

$$T_{3,k} = f(T_{4,k} - \tau + Y_k) + \theta. \tag{2.13}$$

Stacking all these equations for each k into matrix form, one obtains an overdetermined system of linear equations. This type of equations can be solved, for example, using the technique of least squares to be explained in Sect. 3.1.1.

In this way, several measurements can be used to find the most probable offset, slope, and transmission delay for a pair of stationary nodes. In the case that there is no slope between the two nodes or this slope can be neglected, the equations simplify to the equation derived before, and the best estimate of the offset can be given as the mean of the results of the previously described mechanism applied to each of the k measurements individually.

2.3.1.2 One-Way Message Exchange

Opposed to the two-way message exchange system, one can also solve the problem of time synchronization with only one sending reference station and several receiving stations. For the one-way message exchange, Fig. 2.21 depicts the timestamps for which the client has access. This includes a timestamp of the master transmitted inside the packet and a timestamp of the slave created with the local clock of the slave at the time of arrival.

Unfortunately, there is no possibility to estimate the propagation delay from these values. Hence, this synchronization scheme can only be used to set the receiving clock to the timestamp of the sender which includes an offset value as well as the actual transmission delay.

The relevant equation for a single message is very similar to the case of the two-way message exchange and can be given as

$$T_{2,k} = f(T_{1,k} + \tau + X_k) + \theta.$$

Assuming that there is no relevant clock drift, that is, $f \approx 1$, we are left with the following equation:

$$T_{2,k} = T_{1,k} + \tau + X_k + \theta.$$

This equation clarifies again the fact that delay and offset cannot be isolated in this setting: the two unknowns τ and θ cannot be distinguished, and hence, only their sum $\tau + \theta$ can be estimated.

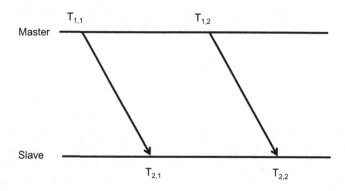

Fig. 2.21 One-way message exchange mechanism

Stacking the equations of several one-way message exchange timestamps leads to a overdetermined set of linear equations, which can, again, be solved by applying least squares regression using the technique explained in Sect. 3.1.1.

The interesting point with this time synchronization scheme is that the receiving stations have a logically correct time with respect to observing packets scheduled in the unknown master time. More distant nodes will have their clocks tuned such that all nodes have the same local time as the master had when transmitting the synchronization packet. This leads to the pleasant situation that a packet sent out at a given time by the master is observed at exactly this time in the local clocks by each slave independent of distance and delay.

2.3.1.3   Receiver–Receiver Synchronization

A very interesting case is the situation, where some infrastructure sends out a message at a specific time. Then different receivers exchange their timestamp of packet arrival and can calculate their relative clock offsets with each other. One of the most important properties of this approach is the fact that the synchronization can be carried out completely independent from the time of the master sending the message. In this way, every wireless infrastructure sending advertisements or other packets receivable by multiple stations can be used to collect initial timestamp data and organize a collective synchronous time for all receivers. The drawback of this method is the fact that the nodes need a method for peer-to-peer exchange of timestamp information. Figure 2.22 depicts this situation.

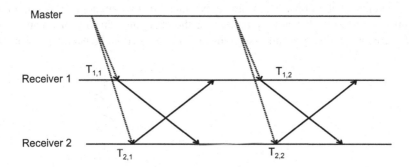

Fig. 2.22 Receiver–Receiver synchronization using a reference signal, which need not contain time information

2.3.2 Acceleration

Acceleration is the rate of change of speed of a mobile entity with respect to its own inertial frame. It is a very interesting quantity for a mobile device as it can be measured directly inside this inertial frame. Unlike speed and location, acceleration has direct impact on the accelerated object's mass.

The main principle of measuring acceleration is given by observing the impact of acceleration onto a test mass mounted flexibly but bound by a spring. In case of acceleration, this mass is changing its place until the spring force cancels the acceleration. Now it suffices to measure the displacement of the mass to find out the actual acceleration.

Figure 2.23 depicts this basic construction. The mass is hanging bound with a spring. When the mass experiences acceleration on the vertical axis, the distance between the mass and the lower line changes and can be measured.

As a consequence from this basic principle of measurement, we have two problems: first of all, the measurement always contains the earth's gravitational pull, and secondly, the measurement principle limits the measurement to one dimension. Therefore, the movement of the mass is often limited to one direction by construction such that only the acceleration in direction of the spring can have impact to the mass. In order to remove the impact of earth acceleration on accelerometer measurements, additional sensors have to be used. One needs to find out the orientation of the accelerometer with respect to earth and can then remove the constant earth acceleration from the accelerometer result.

The most common type of accelerometer nowadays are microelectromechanical systems (MEMS). They have the working principle as described before. However, the elements of the construction are very small. In this situation it is common practice to measure distance by measuring the changing capacity of a capacitor. These changes are very small due to the small construction. Hence, measurement

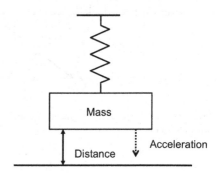

Fig. 2.23 Principle behind acceleration sensors

electronics are directly placed on the MEMS device. Furthermore, three basic accelerometers are packet together on a single device in a pairwise orthogonal manner. The resulting device can then be used to measure acceleration along three independent axes. Accelerometers of this construction are very small and cheap and provide quite good measurement range. They are nowadays built into most smartphone platforms and provide a lot of possibilities for novel applications. From smartphone acceleration, one can decide whether the system is laying around or moving or one is even able to find out the mode of transportation. Very simple applications just assume that the axis with highest acceleration value points towards the ground and can adjust the screen orientation as needed.

2.3.3   Rotation

In order to reliably remove unwanted earth acceleration from acceleration measurements, one needs another sensor system, which can provide exactly such information. However, the actual rotation of a device cannot be observed from within the device in a direct manner. Therefore, rotational acceleration is measured by a gyroscope. The basic idea of a gyroscope is based on the physical law of conservation of the angular momentum: a quickly rotating mass is striving towards not changing its angular momentum. Therefore, when the rotating mass is mounted inside a device such that every rotation of the rotating mass with respect to the device is possible, then any rotation of this device will result in a rotation of the spinning part relative to the construction as this spinning part keeps the same orientation with respect to the surroundings. This rotation of the spinning part in comparison to the housing can then be measured and provides a robust method of detecting rotation. Figure 2.24 depicts a simple construction and the result of rotating the device.

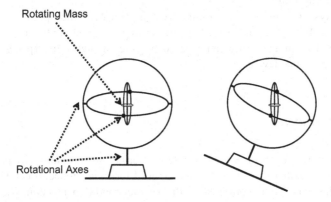

Fig. 2.24 Principle of the gyroscope

The rotating mass in the middle (shaded disk) is bound to a construction of three concentric circles assembled in a way such that they can rotate with respect to each other. Each of these circular constructions is able to rotate around one axis giving the mass complete freedom of rotation. As the rotating part keep its general orientation, the rotation of the circles with respect to each other can be measured and provide rotation information.

2.3.4 Audio and Ultrasonics

For indoor navigation, sound and ultrasonic devices have become an important type of sensor. This is due to the fact that the ambient noise can be captured by a microphone and is sometimes characteristic enough to distinguish between different places. Ultrasonic waves have often been used in positioning appliances due to their welcome properties: humans typically do not hear ultrasonic signals, and a positioning infrastructure can greatly gain from the slow propagation speed as compared to electromagnetic waves. Moreover, reflections of ultrasonic signals often cover semantically equivalent places. An ultrasonic beacon will be easy to detect inside a room, but its signal level will be very low outside the room. In this way, ultrasonic and sound applications are very well suited for representing the notion of place from human perception.

The sensor equipment for both cases is quite similar, and the working principle of a microphone is easily explained: the sound itself induces vibration of some material, and these vibrations can be converted to an electrical signal in various ways. A condenser microphone, for example, contains a thin conductive membrane. This membrane converts sound waves to movement of itself. It forms a capacitor together with another fixed plate, and the vibration changes the capacity of this construction. This capacity can then be used to generate an electrical signal out of the surrounding sound.

Nowadays, smartphones are commonly equipped with microphones, which are capable of collecting sound from the surroundings in good quality for hands-free device use. This functionality can be used to generate audio fingerprints from the surroundings of the device.

2.3.5 Barometer

The barometer is a sensor, which is able to measure air pressure. Air pressure is linked to the weather as well as to the height of location. Hence, it can be used to calculate the one out of the other: given the weather condition, it becomes possible to calculate the height and vice versa. The following formula gives the relation between air pressure and height in standard weather criteria:

$$p(h) = 1013.25 \left(1 - \frac{0.0065h}{288.15\,\mathrm{m}}\right)^{5.255} \mathrm{hPa}.$$

The barometer can also be used to estimate height difference in time. It is quite easy to distinguish between ascending and descending stairs from a barometer, while the distinction of these two movement patterns using acceleration measurements is quite challenging. Modern smartphones are often equipped with barometers as they are relatively inexpensive and allow innovative application scenarios.

2.3.6   Magnetometer and Digital Compass

A magnetometer is used to measure the magnetic flow. Magnetometers are often based on Hall sensors, which use the Hall effect to detect magnetic fields. A hall sensor consists of an electrical conductor with current flowing through the conductor. When the sensor is put into a magnetic field orthogonal to the direction of the current, the electrons are deflected from their straight path. Hence, one side of the conductor becomes negatively charged, and the other side becomes positively charged. This results in a measurable voltage between both sides of the sensor, the Hall voltage. This can be used to detect and measure magnetic fields in one direction. The basic construction of a Hall sensor is depicted in Fig. 2.25.

In order to be able to measure arbitrary electromagnetic fields, it is customary to put three independent hall sensors together, one for each axis. In this way, arbitrary electromagnetic fields can be measured.

From these measurements, it is possible to derive the magnetic north direction using the earth natural magnetic field leading to a working digital compass.

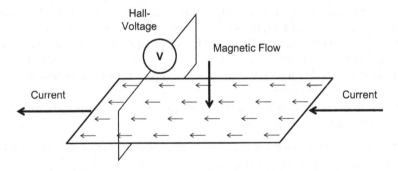

Fig. 2.25 Principle of the Hall sensor

However, the earth magnetic field is relatively weak, and a lot of perturbations can be observed inside buildings. It is not always possible to track a sensible north direction with a magnetic compass alone. Therefore, the gyroscope can provide additional information.

2.3.7   Wireless Infrastructure Components

In mobile computing scenarios, network access is often provided using an infrastructure of fixed network access points. This concept includes Wi-Fi installations as well as cellular networks. The mobile device then needs mechanisms to detect the availability of these infrastructures if not yet connected and to manage handover from one access point to the next in cases, where the mobile device is connected while moving. Therefore, infrastructure components often send out small messages called beacons using a broadcast channel to advertise their availability. Mobile devices then use signal metrics such as the signal strength in order to assess the achievable connection quality and to detect situations in which a handover should be performed. This has two implications: first of all, mobile devices can get identification information from beacons. This is the MAC address in the case of Wi-Fi and a cell identifier and location area code in the case of cellular networks. Moreover, the mobile device can measure the signal conditions in some way in order to select the best serving station. The signal availability can be used to assess some physical proximity between a mobile device and some infrastructure component especially in the case where the beacon contains some uniquely identifying attributes. The signal metrics including RSS can be used to estimate the distance between the mobile device and the infrastructure component or for pattern matching in complex environments.

While Wi-Fi and cellular networks are among the most widely distributed systems, there are a lot of other radio communication systems defined including Bluetooth, ZigBee, and UWB communication. However, these are not yet widely deployed as an infrastructure. One notable exception is iBeacons. These are small devices operating a Bluetooth Low Energy stack and are deployed solely for the purpose of advertising their presence in order to provide proximity-based services based on Bluetooth Low Energy.

Using information of wireless infrastructures has several important advantages making these approaches promising:

- *Availability:* The infrastructure is already available, and exploiting it for location-based services does not introduce new costs.
- *Energy:* The signal metrics have to be collected at any time while using this infrastructure in order to be able to correctly manage handovers and network association. Therefore, no additional energy is wasted in the measurement stage.
- *Passivity:* Often, it is possible to extract signal metrics or proximity information without associating with a network. This is due to the fact that the establishment

of authorization and authentication needs a simple communication channel beforehand for which beacons provide availability information. Moreover, the mobile device often does not need to send any data and, hence, uses only a fraction of the energy as compared to some active sensor system.

In summary, wireless communication systems rely on measuring signal metrics and advertising access point availability. If a wireless communication infrastructure is in use by a mobile device, it is likely that the mobile device already collects measurements. Hence, the exploitation of these measurements does not introduce new energy or infrastructure cost except the amount of energy used for calculation.

2.4 Summary

Wireless communication systems have made location-based services possible in that they allow for both: reliable global position determination as shown with GNSS systems and communication of information. Without wireless communication, smartphones would not have been possible, and personalized services would be limited to users sitting in front of a personal computer with Internet access. However, in buildings, the propagation complexities due to multipath have not yet been successfully mitigated. The physical properties of communication systems provide signals that vary depending on the spatial relation between a sender and a receiver and can, hence, be used to infer about the position of mobile devices. The ubiquitous ability to communicate to large datasets over the Internet allows for information services in which the actual information need not be known in advance but can be generated or extracted from Internet services in real time.

In the next decades, it is to be expected that mobile communication capabilities will grow more and more and that, thereby, the usefulness of mobile services will increase to a scale, where most people will rely upon them in their everyday lives. This makes considerations about privacy and security as important as the research on technical capabilities for inferring about the position and the situation of a mobile user.

References

1. Bahl, P., Padmanabhan, V.N.: Radar: an in-building rf-based user location and tracking system. In: Proceedings of the Nineteenth Annual Joint Conference of the IEEE Computer and Communications Societies (INFOCOM), vol. 2, pp. 775–784 (2000)
2. Geier, J.: SNR-Values and Microsoft Windows Network Status. Online (2012). http://www.wireless-nets.com/resources/tutorials/define_SNR_values.html
3. Hall, M., Frank, E., Holmes, G., Pfahringer, B., Reutemann, P., Witten, I.H.: The WEKA data mining software: an update. SIGKDD Explor. 11, pp. 10–18 (2009)
4. Motley, A., Keenan, J.: Personal communication radio coverage in buildings at 900 MHz and 1700 MHz. Electron. Lett. 24(12), 763–764 (1988)

5. Quinlan, J.: Induction of decision trees. Mach. Learn. **1**(1), 81–106 (1986)
6. Quinlan, J.: C4.5: Programs for Machine Learning.   Morgan Kaufmann Publishers, San Francisco, CA, USA (1993)
7. Rappaport, T.S.: Wireless Communications: Principles and Practices.   Prentice-Hall, New Jersey (2002)
8. Schneier, B.: Applied Cryptography, 2nd edn. Chichester, West Sussex, England, Wiley (1996)
9. Shannon, C.: A mathematical theory of communication. Bell Syst. Tech. J. **27**, 379–423 (1948)
10. Shannon, C.: Communication in the presence of noise. Proc. IRE **37**(1), 10–21 (1949)
11. Stallings, W.: Wireless Communication and Networks. Prentice Hall, New Jersey (2002)
12. Tran-Minh, N., Do-Hong, T.: Application of raytracing technique for predicting average power distribution in indoor environment. In: Second International Conference on Communications and Electronics, pp. 121–125 (2008)
13. Wölfle, G., Wahl, R., Wertz, P., Wildbolz, P., Landstorfer, F.: Dominant path prediction model for indoor scenarios. In: German Microwave Conference (GeMIC) (2005)
14. Zehner, M.L., Bannicke, K., Bill, R.: Positionierungsansätze mittels WLAN-Ausbreitungsmodellen (2005)

Chapter 3
Basic Positioning Techniques

If people do not believe that mathematics is simple, it is only because they do not realize how complicated life is.

John von Neumann

Indoor positioning is the task of inferring the location of a mobile device inside a building. Often, indoor positioning is identified with indoor navigation as many people seem to believe that positioning is the only missing prerequisite to provide guidance services comparable to a GPS-based navigation system outside buildings.

Still, a lot of questions remain open once indoor positioning becomes widely available. This is mainly due to the fact that the semantic structure of a building is much more complex than a road network. Without additional ideas and algorithms, an indoor positioning system would have to provide accuracy below 1 m of expected error. This accuracy is not even available outside buildings.

Another aspect of indoor positioning is the fact that making indoor positioning available inside one building induces the additional challenge of seamless positioning: When should a mobile device switch over to the indoor system? How can the positioning system for the outside space deliver bootstrap information to the indoor positioning system when a user is walking into a building? And what happens, when the user leaves the building?

But before we start to discuss the vast number of possibilities to infer a mobile device inside buildings, let us look at a simple question, which should always be asked before heading for indoor positioning technology:

Why should we introduce navigation inside a building?

Most buildings are either used by a small group of users with varying tasks (e.g., home, bureau, etc.) or by a large group of users with a similar task (e.g., airport, hotel, etc.). In both cases, classical signage or some smart digital signage will be enough to guide the majority of users. Hence, indoor positioning is for different application scenarios and should, hence, not be discussed isolated from a concrete application. This is due to the fact that the most important trade-off before choosing an indoor positioning system is between cost and accuracy. Moreover, the needed accuracy is dictated by the application.

For example, a proximity marketing solution could be provided to users using a positioning technology. However, the application is not at all interested in the

© Springer International Publishing Switzerland 2014
M. Werner, *Indoor Location-Based Services*, DOI 10.1007/978-3-319-10699-1_3

position in the sense of location coordinates but much more into the fact of being near to some point. It can be a good idea in such cases to distribute digital beacons at these places, for example, based on Wi-Fi, Bluetooth, or Radio-Frequency Identification (RFID) technology.

A lot of serious interest in positioning technology stems from security applications, quality control in manufacturing, and safety applications in high-risk areas. In all of these cases, the location of a mobile device is used as an additional feature inside an existing process reducing errors there. For example, quality control is usually a complex process based on education, management, self-control, and motivation. In these applications, the decision between different indoor navigation systems is guided by the need to find out enough information to truly aid the process and, which is often more difficult to achieve, by the invented errors based on wrong location estimates. The critical effect of seldom wrong location measurements is best illustrated in the area of ambient assisted living (AAL). Assume a location system is tracking an elderly person inside his own room and wrongly detects that the person does not move anymore. Then, an ambulance might be sent to the elderly person in error leading to high cost.

Altogether, the central question before deploying any indoor positioning system is the following:

Does the system provide a sufficient advantage in the average case to accept the disadvantages the system will introduce?

An indoor positioning system does not only introduce disadvantages due to wrong location estimates: It is very important to incorporate privacy discussions into the design step of any indoor positioning infrastructure. Unfortunately, for many applications, the advantage of indoor positioning does not directly aid the one whose privacy might be violated. Therefore, a lot of deployments face the problem of communicating the need and getting the systems disadvantages accepted by the users.

The rest of this chapter is organized as follows: Sect. 3.1 introduces the basic algorithms of location determination. These algorithms are applicable inside and outside buildings in the same manner. Section 3.2 explains properties which can be used to compare indoor positioning techniques with each other. Section 3.3 describes a short selection of real-world indoor positioning systems and techniques.

3.1 Methods for Location Determination

As suggested by the previous exposition, there exists a large variety of positioning techniques and positioning systems. However, there is only a limited number of algorithms and methods to infer location information from measurements. Therefore, we will organize this section along these algorithms.

A central problem of inferring location is that this inference is usually based on a set of measurements of physical sizes. And these measurements usually contain a considerable amount of noise or even systematic errors of measurement. For the algorithm of circular lateration, for example, one assumes that the mobile device is located on circles around known locations for which the radius has been measured as the distance between the mobile device and a respective reference location. Due to noise, these circles will merely never intersect in a single point. To successfully deal with this type of problem, the next Sect. 3.1.1 introduces the method of least square estimation. This method enables us to calculate the most probable solution to an overdetermined and possibly inconsistent system of linear equations. This method of least squares is the central tool to enable a wide range of geometric location determination algorithms.

3.1.1 Method of Least Squares

Observing the world by means of measuring physical sizes is generally subject to different classes of errors. First of all, the device used to measure a physical size can introduce errors. Moreover, measurements are often stored and communicated in digital form introducing limited precision. Due to these errors, a reliable observation of the world can often only be achieved by repeated measurements. These repeated measurements will contradict each other, and these contradictions need to be resolved.

Let us assume that there is a linear relationship between the values to be observed and the values actually measured. Then, a sufficient number of measurements leads to an overdetermined linear equation

$$Ax = b,$$

where the vector b contains the actual measurements, the vector x contains the value to be determined, and A expresses the theoretic or expected relationship between both. For this situation, Carl Friedrich Gauss and Adrien Marie Legendre found a method in 1795 merely at the same time and apparently independent from each other to find the most probable value of x.

To remove some subtleties from the discussion, we will assume from now on that the matrix A has maximal rank. In other words, the column vectors of A are linear independent. If this is not the case, redundant columns can be removed without introducing problems. In fact, this is usually performed automatically by computer libraries implementing least squares.

As a perfect solution x with $Ax = b$ does not exist due to measurement errors, the objective is simplified and formulated as follows: find the value x, which minimizes the norm of the error function

$$||r(x)|| = ||b - Ax||.$$

A perfect solution corresponds to a vanishing error function and a norm of zero in this expression. When using the Euclidean norm, we can simplify the exposition by squaring both sides. The norm is the square root of the scalar product. Hence, a minimal norm corresponds to a minimal scalar product. As the norm is defined to be the square root of the scalar product, we can calculate

$$||r(x)||^2 = (b - Ax)^T (b - Ax) = x^T A^T Ax - 2x^t A^t b + b^T b \to \min.$$

We want to find the spot x, where this expression is minimal, and we can just use differential calculus therefore: a necessary condition for a function to reach the minimum is that the first derivative vanishes. The first derivative of the above formula can easily be derived and requiring this first derivate to vanish results in the following equation:

$$2A^t Ax - 2A^t b = 0.$$

This equation is called normal equation of the overdetermined system of linear equations $Ax = b$ and is usually given in the equivalent form:

$$A^t Ax = A^t b. \tag{3.1}$$

It is easy to see that $A^t A$ is a positive semi-definite, symmetric matrix. As a result, the normal equation can be easily solved for x, as this equation is not overdetermined anymore. One can show that this x actually minimizes the error function norm. The most important result justifying the use of this approach, especially the choice of error function, is known as the Gauss–Markov theorem. This proves that the given algorithm provides a best, linear, unbiased estimate (BLUE) of the value x. The preconditions of this theorem, namely, that the errors have zero expectation and equal variance, are usually fulfilled when using similar physical sizes.

For details regarding efficient implementations and numerical stability of results, the interested reader should consult introductory books on numerical mathematics such as [7, 14, 18].

3.1.2 Lateration

Lateration is the process of estimating the location of a mobile device given distance measurements to a set of points with known location. Figure 3.1 depicts this situation. The distance measurements are erroneous and limit the locus of the mobile device to the area between the dashed lines. Hence, a good approximation to a system of equation needs to be found in which the locus of the mobile device is bound to circles with radius given by measurements around the known locations.

Fig. 3.1 Lateration: three
erroneous distance estimates
to base stations of known
location result in an estimate
of the location of the mobile
device

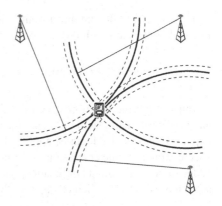

Put formally, the location $p = (x, y)$ of the mobile device must fulfill all equations
k_i describing the circles around the base station locations given as $p_i = (x_i, y_i)$
with radii d_i given by measurements:

$$d_i = k_i(x, y) = \sqrt{(x_i - x)^2 + (y_i - y)^2}\; i = 1 \dots k. \tag{3.2}$$

This is a nonlinear problem, and hence, the least squares algorithm cannot be
applied directly. To overcome this problem, it is customary to use an iterative
approach based on having an initial coarse estimate of the location. Then the Taylor
approximation around this current estimate can be used to linearize the system of
Equations (3.2). This linearization is only valid around the previous location, and
the resulting linear system of equations can be solved using the techniques of least
squares described in the previous section. This linearization is based on the Taylor
expansion given by

$$f(x) = \sum_{i=1 \dots n} \frac{f^{(i)}(x_0)}{i!} (x - x_0)^i + R_{n+1}(x, x_0).$$

The term R collects the remaining error of using a finite sum. To linearize a
system of equations using Taylor expansion, one can set $n = 1$, and for the case
of lateration, one needs the partial derivatives in both directions to construct the
Taylor sum as a vector expression. These partial derivatives of Eq. (3.2) are given as
follows:

$$\frac{\partial}{\partial x} k_i(x, y) = -\frac{x_i - x}{\sqrt{(x_i - x)^2 + (y_i - y)^2}}$$

$$\frac{\partial}{\partial y} k_i(x, y) = -\frac{y_i - y}{\sqrt{(x_i - x)^2 + (y_i - y)^2}}.$$

Let now $(\tilde{x}, \tilde{y})$ denote the current estimate of location. Then using the measurements d_i, one is left with the following system of linear equations:

$$d_i = k_i(\tilde{x}, \tilde{y}) + \frac{\partial}{\partial x}k_i(\tilde{x}, \tilde{y})(x - \tilde{x}) + \frac{\partial}{\partial y}k_i(\tilde{x}, \tilde{y})(y - \tilde{y})\ i = 1 \ldots k. \qquad (3.3)$$

Introducing the notation $\hat{x} = (x - \tilde{x})$ as well as $\hat{y} = (y - \tilde{y})$, this results in an overdetermined system of linear equations for which the method of least squares explained in Sect. 3.1.1 can be applied directly. Applying this technique results in a vector expressing the correction of the current location estimate $(\hat{x}, \hat{y})$. This can be added to the last estimate of location, and the process can be iterated. The following section provides a detailed example of this approach.

Example 3.1 The following table provides the coordinates of the location of four base stations as well as erroneous distance measurements of a mobile entity located at coordinates $(2, 2)$.

Component	Coordinate	Measurement (d_j)
Base station 1	$(x_1, y_1) = (0/0)$	2.92
Base station 2	$(x_2, y_2) = (10/0)$	8.14
Base station 3	$(x_3, y_3) = (15/10)$	15.46
Base station 4	$(x_4, y_4) = (0/12)$	9.89
Mobile device	$(x_e, y_e) = (2/2)$	–
Initial location estimate	$(\tilde{x}, \tilde{y}) = (20, 20)$	–

As a first step, one calculates the following factors:

$$\alpha_i = \sqrt{(x_i - \tilde{x})^2 + (y_i - \tilde{y})^2}\ \text{for}\ i = 1 \ldots 4$$

coming up in the expressions of the partial derivatives in (3.3). From a geometric perspective, these factors provide the distance of the current location estimate to the different base stations. An optimal estimate for the complete problem is reached when $\alpha_i \approx d_i$.

Using the data from the given table, the first step calculates

$$\begin{pmatrix} \alpha_1 \\ \alpha_2 \\ \alpha_3 \\ \alpha_4 \end{pmatrix} = \begin{pmatrix} \sqrt{800} \\ \sqrt{500} \\ \sqrt{125} \\ \sqrt{464} \end{pmatrix} \approx \begin{pmatrix} 28.284 \\ 22.361 \\ 11.180 \\ 21.541 \end{pmatrix}.$$

Using these factors, one can build up the needed linear system as described in Eq. (3.3) consisting of the matrix A given by the rows A_i as follows:

$$A_i = \left(-\frac{(x_i - \tilde{x})}{\alpha_i}, -\frac{(y_i - \tilde{y})}{\alpha_i} \right).$$

For the given example, this results in the following matrix:

$$A = \begin{pmatrix} 0.70711 & 0.70711 \\ 0.44721 & 0.89443 \\ 0.44721 & 0.89443 \\ 0.92848 & 0.37139 \end{pmatrix}.$$

The right-hand side of the linear equation is given by assembling a vector b row by row as

$$b_i = d_i - \alpha_i.$$

For the given example, this results in the following vector:

$$b = \begin{pmatrix} -25.364 \\ -14.221 \\ 4.28 \\ -11.65 \end{pmatrix}.$$

Solving this system of linear equations using least squares approach, the correction vector $(\hat{x}, \hat{y})$ inside the system

$$A \begin{pmatrix} \hat{x} \\ \hat{y} \end{pmatrix} = b$$

has the following concrete value:

$$\begin{pmatrix} \hat{x} \\ \hat{y} \end{pmatrix} = \begin{pmatrix} -18.62235 \\ -0.23377 \end{pmatrix}.$$

The next location estimate is accordingly given by

$$\begin{pmatrix} \tilde{x}_2 \\ \tilde{y}_2 \end{pmatrix} = \begin{pmatrix} \tilde{x} \\ \tilde{y} \end{pmatrix} + \begin{pmatrix} \hat{x} \\ \hat{y} \end{pmatrix} = \begin{pmatrix} 1.3777 \\ 19.7662 \end{pmatrix}.$$

Iterating this approach, one gets results as given in the following table:

Step number	Location estimate	Correction vector	Norm of the residuum
1	(20.00, 20.00)	(−18.62, −0.23)	25.456
2	(1.38, 19.77)	(0.38, −8.82)	17.777
3	(1.75, 10.95)	(−0.38, −8.88)	8.954
4	(1.37, 2.07)	(0.57, 0.09)	0.631
5	(1.94, 2.16)	(−0.01, −0.01)	0.168
6	(1.94, 2.15)	(0.00, 0.00)	0.163
7	(1.94, 2.15)	(−0.00, −0.00)	0.163

Fig. 3.2 The example and
the iterative process of
lateration

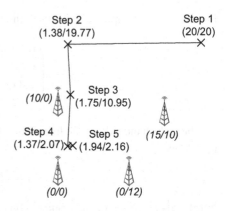

This table shows that the result approaches the true location $(2, 2)$, though it will never reach it due to the inherent errors in the system. The residuum is the distance of the estimate to the true location. From the seventh step, the result does not change inside the printed precision, and hence, the iteration can be stopped. The overall result is then given by the last estimate $(1.94, 2.15)$.

Figure 3.2 depicts the base stations, the starting position, and the iteration steps.

3.1.3 Hyperbolic Lateration

Hyperbolic lateration is a variant of lateration in which the measurement input does not consist of distance estimates to known locations but estimates of distance difference. Assuming that some infrastructure is tightly synchronized and produces events at the same time, which can be received at different times by a mobile station, this is a quite common variant known as time difference of arrival. The most important advantage of this measurement of time differences is that the mobile device does not need to be time synchronized with the sender of a signal. When for two base stations the difference Δd of the distances between the mobile device and both base stations is known, then the mobile entity resides on the hyperbel defined by this distance difference as depicted in Fig. 3.3.

In order to calculate the position, one uses the same approach as used for circular lateration in the previous section. As a first step, the following system of equations is established limiting the location of the mobile device onto the hyperbels defined from the measurements:

$$\Delta d_{ij} = k_i(x, y) - k_j(x, y)$$
$$= \sqrt{(x_i - x)^2 + (y_i - y)^2} - \sqrt{(x_j - x)^2 + (y_j - y)^2} \quad i, j = 1 \dots k, i < j.$$

$$(3.4)$$

Fig. 3.3 Example for hyperbolic lateration: the distance difference between the drawn lines is known fixing the locus of the mobile device onto a hyperbel

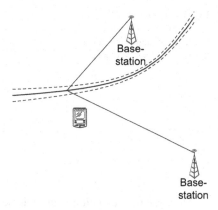

It is customary to only include the distance measurements with respect to one fixed station, say, $i = 1$. This nonlinear, overdetermined system of equations in erroneous measurements is then solved along the same lines as in the previous paragraph. As a first step, the Taylor expansion is calculated and then used to linearize the problem at some estimated location in order to iteratively update these estimations until the process converges.

From a system implementation perspective, the main difference between hyperbolic lateration and lateration is given by which components have to be time synchronized. In lateration using signal propagation delay, each pair of sender and the receiver have to be synchronized. For hyperbolic lateration, the complete set of anchors need to be synchronized, while the mobile device does not need a reference to that time.

3.1.4 Angulation

Angulation is another very common class of positioning approaches in which measured angles between known base stations and mobile devices are used to infer the location of the mobile device. For angulation, there are two general perspectives regarding angles: either the angle between fixed points and mobile devices is measured at those fixed locations or the mobile device measures angles with respect to the incoming signals of base stations. We will discuss only the more common case that the distributed infrastructure measures angles in this section. The other case is, however, very similar. Figure 3.4 depicts an instance of this case.

Ignoring complications such as multipath effects, measurements of angles of incoming radio signals at base stations limit possible locations of the mobile device onto a ray starting at the base station. This can, again, be expressed as a nonlinear system of equations as follows:

$$\alpha_i = \arctan \frac{y_i - y}{x_i - x} \quad i = 1 \ldots k. \tag{3.5}$$

Fig. 3.4 Example of
angulation: three erroneous
estimates of angles provide a
position of a mobile device

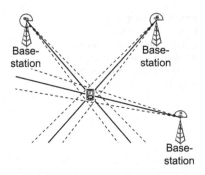

This system of equations can again be linearized by a first-order Taylor expansion leading to an iterative algorithm which refines an initial location estimate by approximating the nonlinear system of equations around these estimates and solving the linearized system there in order to update the location estimate.

3.1.5 Proximity Detection

Proximity detection is a class of location determination algorithms which are based purely on the proximity of the mobile device to previously known locations. The visibility of a Wi-Fi network, for example, results in proximity to the access point as the Wi-Fi signal is limited to a region around the access point. Consequently, proximity detection does not provide location in form of coordinates but rather in form of sets of possible locations. Proximity to a given Wi-Fi access point limits location of the user to a large and complex region. Therefore, proximity to several different locations can be used to intersect these sets and find smaller regions of possible residence of the mobile device. A common simplification is given by assuming that the range of a wireless infrastructure would be well represented by a circle of given radius r. Then proximity results in being located inside this circle, and for several circles, one can limit the possible location to the intersection of the different circles

$$p \in \bigcap_i \{x \text{ mit } \|x - p_i\| < r\}.$$

This simplified situation is depicted in Fig. 3.5.

In extreme cases, proximity detection can be used with very many possible objects and very small radii r such that the location of a device can be tracked down to centimeter accuracy. However, this implies that a large infrastructure is deployed to distinguish between these small areas of location.

Fig. 3.5 Example of
proximity detection

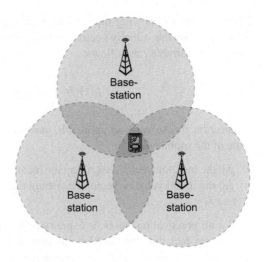

3.1.6 Inertial Navigation

Inertial navigation systems are based on estimating the location of the mobile device
using only measurements made inside the inertial system of the mobile device.
Therefore, the location, speed, and orientation at the starting time are known, and
measurements are used to update this complete movement state. Inertial navigation
is usually based on measuring acceleration and rotation. Sometimes, odometrical
measurements including steps as well as steering angles can also be used.

The most important advantage of inertial navigation lies in the fact that the
mobile device can operate completely autonomous. It does not depend on any
infrastructure. The most important drawback of inertial navigation is that the
location of a device cannot be observed directly from within the inertial frame of
the mobile device. Hence, measurement errors in sensor data will accumulate over
time rendering inertial navigation systems useless after a specific amount of time.

Typical inertial navigation systems are based on using an inertial measurement
unit (IMU) containing six elementary sensors measuring acceleration in three
pairwise orthogonal directions and three gyroscopes each measuring rotation around
one axis. Altogether, these IMUs can be used to update the location and pose of the
mobile device.

Inertial navigation systems in general use the directly measured acceleration
vector $\ddot{p}$ and calculate the current velocity using the following integral:

$$\dot{p}(t) = \int_{t_0}^{t} \ddot{p}(\tau)d\tau + \dot{p}_0.$$

This amounts to knowing the velocity vector at time t. In the same way, one can again integrate the velocity vector to obtain the location function assigning location to the time variable t as follows:

$$p(t) = \int_{t_0}^{t} \dot{p}(\tau)d\tau + p_0.$$

In order to use such an approach, one needs two information about the initial state of the mobile device:

- p_0: the location vector describing the location of the mobile device at time $t = t_0$
- $\dot{p}_0$: the initial velocity vector describing the velocity of the mobile entity at time $t = t_0$

For all practical purposes, it is pretty difficult to use inertial navigation systems due to the fact that the doubled integration is also applied to the error vectors inherently contained in the measurements of acceleration. These can quickly accumulate to a wrong movement state and over time to arbitrary large errors in position. If, for example, the mobile device is accelerated and then again stopped, the errors will seldom cancel and the estimated speed will be nonzero; hence, the estimated location drifts away from the actual location. This is even more problematic, when the earth acceleration has to be removed from the accelerometer readings based on other erroneous measurements.

Due to these problems, inertial navigation is often used only together with another positioning system able to recalibrate the movement state of the mobile device. One prominent example of this is given by inertial navigation using foot-mounted IMUs. In these cases, it is possible to reset the complete movement state to zero during the time where the foot remains in contact with the floor. These zero velocity updates (ZUPT) can lead to very good overall system performance as the errors can only accumulate during the short time frame in which a foot moves freely in air.

3.1.7 Fingerprinting

The previously described approaches are all based on observing a known physical relation between some measurable size and location. In contrast to that, a lot of approaches exist, which do not rely on any such relation but are rather based on reproducibility of patterns of measurable variables. In this way, the more complex an environment and the behavior of underlying physics becomes the more difficult can the physical laws be used to infer the location. However, these complexities make data locally unique and distinguishable leading to a new technique of location determination known as fingerprinting.

The set of measurements at a specific location is similar to the set of measurements taken at the same location at another time or with another device, but not

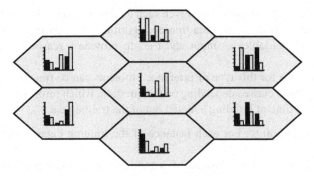

Fig. 3.6 Example of using scene analysis techniques

#	Observable Variables							Location	
1									
2									
3									
4									
5									
6									
7									

Instances

Fig. 3.7 Data mining problems represented by a table

similar to measurements taken at another place. Figure 3.6 illustrates this concept using a cell subdivision of an area and assigning a histogram describing each location cell. This histogram could be given by signal strength of GSM base stations or similar location-dependent measurements.

In general, fingerprinting is a class of algorithms, which are often formulated as classification or regression problems in the form of a data mining problem. These are typically based on two phases: a training phase, in which relations between measurements, coordinates, and labels are collected and stored in a training dataset. Such a dataset is typically a table of measured variables (or variables directly calculated from atomic measurements) and location. Figure 3.7 depicts this situation. The rows of the table are called instances and combine a set of data at a given situation. In the case of localization, this consists of all measurements made at one fixed location called observable variables in the figure, as well as a description of location given by a label or coordinates. When several parameters cannot be observed in one instance, the cells can be marked as unknown. Some data mining schemes can perfectly deal with missing values in both the observable variables and the label or location.

A data mining algorithm is now given such a table including the rows describing location. Using this data, the data mining algorithm takes an instance containing only observable variables as input and tries to provide a reasonable estimate of location.

The algorithms for this type of inference problems can be roughly grouped into the following three classes depending on the question, which information is actually available at the time of building a model out of the training data:

- *Supervised Learning*: For each instance of the training dataset, the location is given.
- *Unsupervised Learning*: Instead of searching for a model predicting location out of observable variables, unsupervised learning tries to find relations between attributes of instances.
- *Semi-supervised Learning*: Both approaches are combined—a prediction model is built from the training data containing location, and unsupervised learning is being used to enhance this model further.

In general, one must take care that the data mining model generated from a training set does not describe the training dataset too closely. Because then, the model will have difficulties to deal with unknown instances. Hence, the success rate on the training set will be rather high, while the model has problems in generalizing the extracted knowledge to unknown situations. This problem is very common and called *overfitting*. Overfitting becomes easily possible with small training datasets in which the randomness of noise influences cannot be observed and the actual noise sizes are used during classification.

For the case of classification, that is, assigning a label to an unlabeled instance using a training set of labeled instances, a simple and well-known method is given by *decision trees*. A decision tree is usually defined to be a binary tree in which every inner node including the root node contains a test comparing observable variables. For an unknown instance, these tests can be performed, and depending on the result, the left or ride subtree can be expanded until one reaches a leaf node. The leaf nodes then contain the label to be predicted. From a geometrical point of view, the set of training instances consists of points in a high-dimensional space, one dimension for each attribute. A decision tree is then usually constructed in a way such that each level splits the set of training instances into two similar parts. A very classical definition of how to split is the amount of information one gains with a split. The well-known ID3 algorithm splits on single attributes using a threshold. That is the comparisons inside the inner nodes are of the type $a < t_a$ for an attribute a and a threshold t_a. Instances with attribute value smaller than the threshold are descending the left subtree and instances with attribute value higher than the threshold descend the right tree. The split is constructed using the information gain that is a measure about how informative a split was. It is given as the difference between the entropy (see Sect. 2.1.2) of the current training set and the entropy of the set after splitting it using the current attribute. Let therefore A denote some attribute; S denote the current set of training instances; T be the splitting of S using A, that is, $S = \cup_{t \in T} t$; and $p(t)$ the relative count of elements of $t \in T$ with respect to T. Then let H

denote the entropy. The information gain of an attribute together with a splitting T using this attribute is given by

$$\mathbf{InfoGain}(\text{attribute}) = H(S) - \sum_{t \in T} p(t) H(t).$$

The ID3 algorithm now uses the attribute, which provides the highest information gain.

Another fundamentally different method of learning from examples is based on the famous Bayes theorem:

$$P(I|M) = \frac{P(M|I)P(I)}{P(M)}. \tag{3.6}$$

This equation about probabilities means that the probability $P(I|M)$ that an event I happened when M has been measured can be calculated from the right-hand factors. From a practical point of view, this means that we can predict probabilities of events from measurements, when we are able to fill in probabilities to the right-hand side. The right-hand side contains the probability $P(M|I)$ that a measurement M has been made in cases where the event I actually happened and can easily be calculated from a training set. Note that a training set contains measurements associated to events. This has to be multiplied by the overall probability of event I happening. This again can be easily estimated from relative fractions calculated over a training set. This product is then to be divided by $P(M)$, a probability to get measurement result M in general.

Let, for example, I describe the location given by location labels and let M describe a measurement of signal strength of some Wi-Fi access points. Then $P(I|M)$ can be calculated for each location I and gives the probability of being at location I. The right-hand values can be estimated from the training dataset as explained before. Note that the denominator $P(M)$ of Eq. (3.6) can be safely ignored as it does not vary for a single measurement. It suffices to calculate the nominator of Eq. 3.6 for every location I and normalize the resulting values such that they sum up to one.

The *naive Bayes* algorithm makes an additional assumption. It assumes that the attributes are pairwise statistically independent. Let the attributes be named $a_1, \ldots, a_n$. Then, under the assumption of independence, the right-hand factor $P(M|I)$ can be further simplified to

$$P(M|I) = \prod_{i=1\ldots n} P(a_i, I),$$

which makes the calculation of relative counts possible on a per attribute basis in the dataset. This algorithm is called naïve as the assumption of independence is often not justified in practice. However, this algorithm still provides good results in many cases, and in other cases, one can try to automatically reduce redundancy of attributes beforehand such that the input attributes to the naïve Bayes algorithm are more or less independent.

When data mining approaches are used to infer the location of a mobile device inside buildings, they are usually called *fingerprinting* algorithms. The seminal work RADAR [1] can be seen as one of the first working indoor positioning systems based on Wi-Fi signal strength uses a weighted k next neighbor (kNN) approach to estimate the location from incoming signal strength. Therefore, for an instance containing signal strength information, the k next neighbors with respect to signal strength vectors are searched within the training database. Their distance in signal space to the incoming signal strength vector is then taken to calculate the location result for the incoming signal strength vector as the weighted sum of the location of the k next neighbor data points from the training set.

3.2 Properties and Evaluation of Positioning Systems

The evaluation of positioning systems is often based on experiments in which a specific error measure of location is used in order to compare the performance of the positioning system. Before using such a measure, however, one should always divide the possible errors into two parts and evaluate them independently, if possible. The quality of a positioning system can be good or bad in two different dimensions as depicted in Fig. 3.8. The two different properties of a positioning system are called precision and accuracy. Though they have a similar meaning in everyday life, they denote quite different aspects for positioning systems. *Precision* measures the deviation of location estimates for the same location from each other, while *accuracy* measures the deviation from the truth. A very accurate system can be used for long-term location determination where the precision is not too relevant and errors cancel out over time. On the opposite, a precise but inaccurate system can be used to guide local decision or find proximity between two mobile devices, while it cannot be used to provide a reliable link to map information. In optimal cases, positioning systems are precise and accurate at the same time.

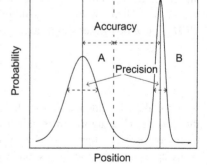

Fig. 3.8 Accuracy vs. precision for a one-dimensional positioning system

However, this is—in general—impossible. Another relevant property of positioning systems is the *spatial resolution* which is defined to be the minimal change of position of the mobile device that can be detected by the system. For proximity systems, this spatial resolution is rather large, while it can be quite small for systems based on lateration. Furthermore, the *temporal resolution* of a positioning system can be of interest: the minimum time for which a position must have changed before the position change can be detected. Again, if a proximity system is based on observing advertisements of a wireless infrastructure and these advertisements are only sent out once per minute, then it takes 1 min until the position can be updated. Furthermore, positioning algorithms can *lag* in time. The lag describes the time difference between an instance of having a specific position in reality and the instance of being informed by the positioning system about this location.

In numerical experiments, one has to carefully select statistical values which can indicate the given properties. However, it is not easy to choose the right ones. It has become common practice in the indoor positioning domain to report the mean of the deviation from the truth as an indicator of accuracy and the standard deviation or variance as an indicator for precision. Using the mean of the positioning error is very illustrative as it is a good estimate of the expectation of error: a user of the system should expect this amount of error for the next measurement. In other domains, especially with more emphasis on the theoretical background, the root mean squared error is used. This is due to the fact that the quality can then be measured globally by comparing it to the Cramer–Rao bound. The Cramer–Rao bound states that the variance of an unbiased estimator is at least as high as the inverse of the Fisher information matrix. Furthermore, an unbiased estimator which achieves this bound minimizes the mean squared error among all unbiased methods and is hence a minimum-variance unbiased estimator. Therefore, all unbiased estimators which try to minimize variance that is increase precision can be compared to this theoretical bound.

Figure 3.8 depicts the estimation of a single location coordinate by two different positioning systems of different accuracy and precision. In this figure, the precision amounts to the spread of the location estimate. The system A is clearly less precise as compared to the system B. The accuracy is given by the distance between the truth depicted as a dashed line in the middle and the estimate of both systems. Clearly, the system B is less precise as compared to the system A.

Several non-numerical aspects are also important in describing the properties of positioning systems. One of the most important properties is *mobility*. Does the positioning system provide position estimates to a device on the move or does it provide location only to devices with limited mobility such as devices inside a specific area of coverage. Furthermore, the *scaling variables* can differ between different positioning systems: for systems with few mobile devices, the price of a high-quality inertial navigation system can be justified. Think of a robot exploring mars as an example. When, however, the number of mobile devices is very high as compared to the area in which they must be located, then it might be better to install some expensive positioning infrastructure and provide the mobile devices with cheap devices such as RFID tags or UWB beacons.

With respect to cost, scaling, and privacy properties of a positioning system, it is common practice to classify positioning systems into the following three classes depending on their working principle [12]:

- *Terminal-based positioning* in which the mobile device calculates position without depending on some infrastructure.
- *Terminal-assisted positioning* in which the positioning is distributed between infrastructure elements and the mobile device.
- *Terminal-free positioning* in which a mobile device is located while the mobile device is passive.

At first sight, the class of terminal-assisted positioning can be a bit confusing, and it might make sense to further subdivide this class. GPS, for example, is clearly a terminal-assisted positioning system. However, the GPS infrastructure does not learn any location information about the mobile devices. It rather provides data from which the device can locally calculate position anywhere on earth. In other terminal-assisted scenarios, the mobile device is used to measure some information, while the infrastructure actually calculates and hence knows the position of the mobile device.

3.3 Examples of Positioning Systems

Inside buildings it is often difficult to estimate the distance between two fixed points due to multipath effects. When trying to measure the distance between two points which are not in line of sight, one estimates the length of the transmission path, which can be quite different from the distance between the two points. Therefore, the lateration approaches often lead to imprecise results unless they are under in line-of-sight conditions. For the measurement of distances, there are two general types of approaches: one is based on signal timing. The distance is then defined by the propagation speed of the signal and the time between sending and receiving a signal. On the opposite, one can use signal strength measurements to estimate distance using a model for the signal strength.

For the timing approach, one considers three general classes depending on which time to measure, namely:

- *Time of Arrival:* The absolute point in time at which some signal (e.g., light, sound, radio) set out at some known place and time reaches the mobile device can be measured.
- *Time Difference of Arrival:* The time difference between two signals sent out from different places at the same time can be measured by the mobile device.
- *Roundtrip Time of Flight:* The time difference between sending out a signal and receiving a reflection of the same signal is measured. Note that reflections can be reflections of the signal in the physical sense or messages sent in response to the package such as ACK frames.

The distances estimated using a time-of-arrival approach together with the propagation speed of the signal can be together with the algorithm of lateration to estimate a location. The most important drawback of this type of positioning system is the fact that the infrastructure as well as the mobile device needs to be tightly time synchronized. If, for example, the clock error at the mobile device is 1 μs and the system uses radio communication for positioning, this time error introduces a length estimation error of

$$s = c_0 t = 300{,}000{,}000 \, \text{m/s} \cdot 0.000001 \, \text{s} = 300 \, \text{m}.$$

Using audio signals such as ultrasonics with the same type of approach leads to a much better localization estimation due to the slow propagation speed of sound of approximately 343 m/s:

$$s = c_0 t = 343 \, \text{m/s} \cdot 0.000001 \, \text{s} = 3.43 \, \text{mm}.$$

A very classical positioning system based on time of arrival is GPS, and the most challenging part of GPS is to let the satellites have a consistent clock. Therefore, each GPS satellites is equipped with an atomic clock, and the first phase of GPS localization is to get the mobile device tightly time synchronized to the satellites.

In many cases, it is infeasible to let the mobile device be synchronized with some infrastructure. One reason for that is given by the problem that time synchronization algorithms in computer networks are often based on timing of messages which change for mobile devices. For time-difference-of-arrival positioning, the mobile device does not need to be synchronized with any infrastructure as it uses its local clock only to measure time differences independent from a global time. This is especially useful in cases where the infrastructure is already time synchronized for other reasons. In cellular phone networks, time synchronization of base stations is often given and is relatively precise as the stations often use GPS for time synchronization. Then it is easy to let the infrastructure send out pilot signals at the same time, and a mobile device just measures time offsets between the reception of these messages.

Roundtrip time of flight-based positioning systems do not need any time synchronization which is their central strength. However, the drawback of this setting is that the timing information need not be correct when non-physical reflections are being used. The same influences and uncertainties occur as discussed together with the two-way message exchange time synchronization in Sect. 2.3.1.1.

The following section collects several typical indoor positioning systems for each of the previously explained approaches. The reader should also consult classical surveys such as [9, 10].

3.3.1 Pseudolites and High Sensitivity GNSS

The radio signals of different satellite-based positioning systems including GPS, GLONASS, and Galileo suffer very much from path loss and multipath effects inside buildings. The received signal strengths of GPS signals are often smaller than the sensitivity of typical GPS receivers. Moreover the assumption that the signal travels a direct line between the satellite and the position on earth is often wrong. The receiver often receives reflections and, thus, estimates wrong distances to the different satellites due to differences between the direct distance and the length of the propagation path of the signal.

In order to use satellite positioning systems inside buildings, these two problems need to be addressed. One idea is to produce highly sensitive signal receivers. If the sensitivity is high enough and the path loss is small enough, the line-of-sight signal could be received as it will be the first signal though not the strongest one. However, in large buildings with many of floors above a receiver, the line-of-sight signal cannot be received even by highly sensitive receivers. Therefore, this approach is applicable in halls with simple roofs. The most prevalent application domain is logistics where a coarse position with accuracy of about 10 m can easily suffice to locate a fork shift inside a hall.

Another approach to enabling GNSS positioning inside buildings is by providing the signal with a fake infrastructure of pseudolites. A pseudolite emulates the signals a mobile device would receive outside from a satellite. Pseudolites provide quite accurate positioning inside a limited area, where the mobile device has free line of sight to the pseudolites. However, they are also very expensive due to the inherent sensitivity of GPS with respect to time delays. In complex multipath environments, pseudolites can only be used inside regions with line-of-sight conditions between the pseudolites and the mobile devices. Hence, for a complete coverage of a complex building, a lot of pseudolites are needed [3].

Another example of extending the coverage of GNSS systems inside buildings is given by the Locata system. This system is based on a time-synchronized network of base stations inside buildings sending signals similar to GPS but inside the license-free ISM band. These signals can then be used by special devices together with GPS signals to infer the position of a mobile device both outside and inside a building with very high accuracy [17].

Another approach of extending GNSS coverage into buildings is given by modeling all complexities of signal propagation. Therefore, very detailed three-dimensional models of buildings including building material and furniture can be used to calculate the expected dominant propagation path of the satellites. If these models are sufficiently correct, then high sensitivity GNSS receivers can be used without the assumption of direct line-of-sight connection to the satellites increasing accuracy [11, 21]. Though these approaches are promising as there is no new infrastructure as, for example, with the Locata system, the creation of sufficient models of buildings is very expensive, and positioning will still be limited to areas, where a high sensitivity GNSS receiver is able to detect the signals and the building model is sufficiently correct.

3.3.2 Light-Based Systems

Using special light-based techniques for indoor positioning is often motivated from the fact that most building material can reflect light. Moreover, these reflections are more deterministic as compared to other signals, and due to the high frequency of the radio waves, pulses can be quite short allowing for highly accurate line-of-sight ranging. The most prevalent systems use roundtrip time-of-flight approach together with physical reflection of modulated light waves. The modulation is only used to distinguish between scattered light and the reflection. These systems are often called *LiDAR systems (light detection and ranging)*. This name has been chosen for its similarity to RADAR. For RADAR systems, radio signal reflections are detected very similar to LiDAR but with higher range. A very typical LiDAR-based system is presented in the work [20] in which LiDAR depth information is used together with an inertial sensor system to derive position. As a natural extension, LiDAR systems can be used to generate maps of the surroundings using techniques of simultaneous localization and mapping (SLAM). A depth image similar to the depth matrix generated by LiDAR systems can be generated in a simpler fashion with limited range and accuracy. The Microsoft Kinect, for example, sends out a pattern of light. The reflections of this pattern are recollected using a camera. From the distortions of the light pattern, a distance is calculated for each pixel of the camera detecting the light pattern. Unfortunately, the range of such an approach is quite limited due to the resolution needed for the camera and the intensity of light needed to make distant patterns detectable to the camera.

A completely different approach to light-based positioning systems is given by light-based proximity detection. Therefore, a modulated light signal is sent out by a mobile device containing an identification of the mobile device. This signal is then captured by a sensor network, and the location of the sensors detecting the signal can be used to infer the location of the mobile device. One of the first systems of this type is the Active Badge system. In the Active Badge system, sensors are mounted to the ceiling detecting infrared signals from badges placed on the shoulder. Of course, this system can also be inverted by making the mobile device detect light signals sent out from a distributed infrastructure. However, for Active Badge, the badges do not need any capabilities except simple light modulation, whereas the other case is better suited for mobile devices with computational capabilities.

3.3.3 Camera-Based Systems

Camera-based systems aim to extract location and movement information out of the same information as a human being out of his visual sense. This approach is promising as it is known that human orientation is mainly based on visual information. However, we are not yet able to reach the same accuracy of orientation using camera systems.

In general, there are two possible deployments for a camera-based positioning system: either the camera is given to the mobile device and location of the mobile device is extracted from the point of view of the mobile device or the camera is mounted to the building and movement information is extracted from the location of a person or object inside the camera stream.

The former approach is often treated similar to the problem of scene analysis. The challenge is to assign location to a camera image. The localization part of the latter approach is often quite easy; however, the correct identification of mobile objects is challenging.

For a mobile camera system, some information is typically extracted from the camera pictures including landmarks, feature points, or geometric peculiarities. These are then compared to a database of these features referenced to location. In some systems, specific landmarks with high probability of reidentification are observed. Some approaches put synthetic landmarks such as barcodes into the environment, while others try to find natural, distinctive landmarks.

A third class of camera-based positioning systems consists of systems trying to extract the camera egomotion out of a sequence of images. Therefore, techniques such as optical flow extraction or SLAM can be used.

Some extraordinary camera-based positioning systems use the camera for example for monitoring the floor space comparably to an optical computer mouse [5, 15].

For the class of stationary camera systems, the extraction of location is relatively simple. An empty scene has to be recorded as a ground truth. Significant changes to this base scene are given by mobile objects; their depth can often be extracted from the X–Y coordinate inside the picture assuming that the object moves on a planar floor. The challenging problem here is the identification of a mobile object in order to assign a location not only to an observation but to an individual. Linked to this problem is also the reidentification of mobile objects across different cameras. A camera can only cover a fraction of a building, and hence, objects can enter the field of view of a camera and leave it again. The question is now how to reidentify objects leaving one camera's field of view and entering another camera's field of view. Typical areas of application are process observation applications where the actual reidentification of objects is not too important. In these cases, the camera system differentiates between moving objects inside the field of view and tries to detect noticeable objects moving differently from the majority of objects.

3.3.4 Radio-Based Systems

Today, most positioning systems inside buildings rely on radio technology. With radio technology, it is possible to reach extremely high accuracy. Moreover, radio technology has seen large-scale deployment resulting in relatively cheap radio hardware. Additionally, radio communication infrastructures are basically everywhere. GPS reaches the whole surface of the earth, Wi-Fi enables location

Fig. 3.9 Geometry of an
antenna array

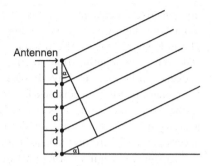

awareness without using GPS, and cell tower signals of cellular networks provide
another wide-spread infrastructure of radio communication systems.

Positioning based on radio signals can be based on signal strength information
as the signal energy decreases with distance. Other systems are based on accurate
timing information. With ultra-wideband (UWB) signals, it is possible to transmit
and detect very short pulses allowing very precise positioning due to the precise
calculation of signal delays. Inside buildings, however, the length of the propagation
path is not always a good indicator for the distance between the sender and the
receiver. Thus, these systems are often limited to line-of-sight conditions, and a lot
of infrastructure is needed to cover several rooms and floors.

A third class of radio positioning systems is based on angle estimation. An array
of antennas can be used to determine the angle from which a radio signal has been
sent out. Alternatively, the same antenna array can be used to transmit radio signals
into a specific direction. If the sender and receiver are far enough from each other,
it is admissible to model the radio waves propagation direction as parallel lines as
depicted in Fig. 3.9. In this case, the time delay of the same signal (e.g., the phase
difference) at those different antennas can be measured and results in an estimate of
the angle. In the opposite case, the signal can be sent out into a specific direction by
choosing appropriate delays for the different antennas. The superposition of waves
is then maximally constructive into the direction given by the angle α. The situation
depicted in Fig. 3.9 leads to the following equation relating the angle α to the time
delay Δt:

$$\Delta t = \frac{d}{c_0} \sin \alpha.$$

Finally, some radio-based systems use the complexity of signal propagation
inside buildings and try to collect reproducible features changing with location.
A very classical example for this type of localization is RADAR [1]. RADAR
collects Wi-Fi signal strength together with location in a training phase. The
localization is then performed by assigning to incoming signal strength information
the location of the most similar database entries. It has been shown in experiments
that this information is actually characterizing location as long as there is not
too much change to the building. In practice, systems based on fingerprinting

are sensitive to changes in the surroundings and need regular recalibration. The main advantage of using Wi-Fi fingerprinting is, however, that an existing and ubiquitously available infrastructure can be used. Wi-Fi access points, in general, send out their identifications regularly, and the signal strength of these beacon packets can be measured by all Wi-Fi devices as this signal strength is also used to manage handover decisions for Wi-Fi networks with multiple access points. Furthermore, mobile devices regularly send out scan requests to the surrounding access points in order to quickly detect handover situations and to discover hidden access points. Therefore, the infrastructure of Wi-Fi access points is also able to collect signal strength information of mobile devices in many cases.

Another radio technology capable of positioning devices is given by *RFID* deployments. RFID systems are based on two components: RFID readers and RFID tags. The tags are small and cheap electronic components often only able to store and communicate an integer number used for identification of mobile objects. These tags can be either passive or active. Passive tags do not need a current supply and are powered by the RFID reader using induction. Active tags possess their own power supply and can be used more flexibly. The range of RFID systems is typically very limited making proximity-based positioning possible. Therefore, RFID readers are commonly placed in sensible locations such as doors and can be used to split the navigation space into zones. For each tag attached to some mobile object, the zone of residence is then known.

3.3.5   Inertial Navigation

Inertial navigation systems subsume all navigation systems based on measuring changes of the inertial system of the mobile device. As these changes reflect only relative measures, no absolute position can be calculated. Sensors for this type of navigation systems include accelerometer, gyroscopes, odometers, and magnetometers. In this area, only few and very specialized systems for the indoor area have been successfully demonstrated. This is due to the inherent inaccuracy of inertial navigation by errors accumulating over time. Therefore, the inertial navigation system needs external support. Sometimes, this can be provided from another positioning system; sometimes this is just given by points in time, where it is possible to reset the movement state to zero. These zero-update systems often use foot-mounted IMUs and reset movement to zero during the phase where the foot is fixed to the ground. It is also possible to integrate map information into the calculation of position using particle filters. Due to the quick change of sensor quality and the general interest in inertial navigation systems, a lot of work has been done in the last years including [4, 6, 8, 13, 19, 23, 24].

The area of integrated solutions consisting of some coordinate-based positioning system with possibly slow update rates aided with a high-update inertial navigation system between position fixes from the coordinate system is very promising and can lead to an overall system with high-update rate and limited error.

3.3.6 Audio-Based Systems

Audio-based systems use the propagation of audio waves in space in order to locate a mobile device. Especially with sound, several different physical effects can be used to infer the location of the mobile device. Simple audio-based systems provide ultrasonic signals not disturbing humans to identify locations and provide proximity-based awareness similar to the Active Badge system. Due to this similarity, one of the classical such systems is called Active Bat system [22]. This system is based on a sensor network of ultrasonic microphones and detects mobile devices sending out ultrasonic identifications up to centimeter accuracy. The propagation of sound inside buildings is very natural and often better than the propagation of light signals. A lot of building material reflects and scatters sound. In this way, semantic places such as rooms or hallways can easily be filled with the same sound identification, which is not spreading to adjacent rooms too much.

Due to the slow propagation speed of audio (approx. 343 m/s), it is possible to use several microphones to detect the angle out of which a specific audio signal has been received with great accuracy. This can even be used to localize the origin of a shoot in military applications [2].

Another approach to using audio information for indoor localization is given by recording the typical ambient noise of different rooms inside a building. Approaches of this type are quite similar to Wi-Fi fingerprinting in that they take an audio signal of some fixed duration and calculate a fingerprint out of this sound signal in a manner such that the fingerprint is characteristic for location: it changes across different locations while it is reproducible at a fixed location over time. This type of positioning system is, of course, only applicable in cases where a typical ambient noise exists and varies sufficiently between locations. The most promising application domain is thus given by industrial environments.

3.3.7 Pressure-Based Systems

An exceptional positioning system called Smartfloor is based on distributing pressure sensors across the floor measuring the presence of objects by their pressure induced against ground [16]. The main advantage of such a system is its unobtrusiveness. The localized objects do not need to interact in any non-natural way with the environment. Just as with camera infrastructures, the main problem with such an approach is with respect to identification and reidentification of objects inside crowded scenes. This approach is a very promising approach for applications of AAL as these overcrowded scenarios are less likely to occur and the risk for a patient is low when many other people are around. For these applications, the unobtrusiveness is the most important factor in order to allow a system to monitor the patient movement at any time.

Table 3.1 Most important characteristics of positioning approaches

Algorithm	Input sizes	Limitations
Lateration	Length, distance, time	Time synchronization, multipath
Hyperbolic lateration	Length differences, delays	Only infrastructure needs to be time synchronized
Angulation	Angles, phase differences	Multipath
Proximity detection	Visibility, physical proximity	Simple and reliable, often only coarse location
Inertial navigation	Acceleration, rotation, movement	Errors accumulate Independent from infrastructure
Fingerprinting	Feature vectors	Stable with respect to multipath and complexities, sensitive to (small) changes in the environment
SLAM	Inertial navigation, depth images	Often accurate, independent from infrastructure, computationally challenging

3.4 Summary

As described in this chapter, the available technologies and algorithms for detecting the position of a mobile device are manifold. The first part of this chapter introduced the main algorithms for determining position from measurements, while the second part introduced actual examples for organizing along the measurable variables. A very short summary of algorithms is given in Table 3.1 in which the most important advantages, limitations, or disadvantages of different schemes are mentioned.

References

1. Bahl, P., Padmanabhan, V.N.: Radar: an in-building rf-based user location and tracking system. In: Proceedings of the Nineteenth Annual Joint Conference of the IEEE Computer and Communications Societies (INFOCOM), vol. 2, pp. 775–784 (2000)
2. Boggs, J.: Geolocation of an audio source in a multipath environment using time-of-arrival. Tech. rep., DTIC Document (2004)
3. Cobb, H.S.: Gps pseudolites: theory, design, and applications. Ph.D. thesis, Stanford University (1997)
4. Davidson, P., Collin, J., Takala, J.: Application of particle filters for indoor positioning using floor plans. In: Ubiquitous Positioning, Indoor Navigation, and Location Based Service. IEEE, New York (2010)
5. Dille, M., Grocholsky, B., Singh, S.: Outdoor downward-facing optical flow odometry with commodity sensors. In: Field and Service Robotics, pp. 183–193. Springer, Berlin (2010)

6. Evennou, F., Marx, F.: Advanced integration of wifi and inertial navigation systems for indoor mobile positioning. Hindawi Publishing Corporation EURASIP J. Appl. Signal Process, **2006**, 1–11 (2006). doi:10.1155/ASP/2006/86706

7. Freund, R.W., Hoppe, R.H.W.: Stoer/Bulirsch: Numerische Mathematik 1. Zehnte, neu bearbeitete Auflage. Springer, Berlin/Heidelberg (2007)

8. Goyal, P., Ribeiro, V., Saran, H., Kumar, A.: Strap-down pedestrian dead-reckoning system. In: Proceedings of the International Conference on Indoor Positioning and Indoor Navigation. IEEE, New York (2011)

9. Hightower, J., Borriello, G.: Location systems for ubiquitous computing. Computer **34**(8), 57–66 (2001)

10. Huang, H., Gartner, G.: A survey of mobile indoor navigation systems. In: Cartography in Central and Eastern Europe, pp. 305–319. Springer, Berlin/Heidelberg (2010)

11. Kee, C., Yun, D., Jun, H., Parkinson, B., Pullen, S., Lagenstein, T.: Centimeter-accuracy indoor navigation using GPS-like pseudolites. In: GPSWorld (2001)

12. Küpper, A.: Location-Based Services: Fundamentals and Operation. Wiley, New York (2005)

13. Link, J.A.B., Smith, P., Viol, N., Wehrle, K.: FootPath: accurate map-based indoor navigation using smartphones. In: Proceedings of the International Conference on Indoor Positioning and Indoor Navigation (2011)

14. Ludwig, R.: Methoden der Fehler- und Ausgleichsrechnung. Deutscher Verlag der Wissenschaften, Berlin (1971)

15. Nagatani, K., Tachibana, S., Sofne, M., Tanaka, Y.: Improvement of odometry for omni-directional vehicle using optical flow information. In: Proceedings of the 2000 IEEE/RSJ International Conference on Intelligent Robots and Systems, vol. 1, pp. 468–473 (2000)

16. Orr, R., Abowd, G.: The smart floor: a mechanism for natural user identification and tracking. In: CHI: Extended Abstracts on Human Factors in Computing Systems, pp. 275–276. ACM, New York (2000)

17. Rizos, C., Roberts, G., Barnes, J., Gambale, N.: Locata: a new high accuracy indoor positioning system. In: Proceedings of the International Conference on Indoor Positioning and Indoor Navigation (2010)

18. Schwarz, H.R., Köckler, N.: Numerische Mathematik. Fünfte Auflage, B.G. Teubner-Verlag, Wiesbaden (2004)

19. Storms, W., Shockley, J., Raquet, J.: Magnetic field navigation in an indoor environment. In: Ubiquitous Positioning, Indoor Navigation, and Location Based Service. IEEE, New York (2010)

20. Travis, W., Simmons, A., Bevly, D.: Corridor navigation with a lidar/ins kalman filter solution. In: Intelligent Vehicles Symposium, pp. 343–348. IEEE, New York (2005)

21. van Diggelen, F.: Indoor GPS theory & implementation. In: Position, Location and Navigation Symposium, PLANS, pp. 240–247 (2002)

22. Ward, A., Jones, A., Hopper, A.: A new location technique for the active office. IEEE Pers. Commun. **4**(5), 42–47 (1997)

23. Woodman, O., Harle, R.: Pedestrian localisation for indoor environments. In: Proceedings of the 10th International Conference on Ubiquitous Computing, pp. 114–123. ACM, New York (2008)

24. Xiao, W., Ni, W., Toh, Y.: Integrated wi-fi fingerprinting and inertial sensing for indoor positioning. In: International Conference on Indoor Positioning and Indoor Navigation. IEEE, New York (2011)

Chapter 4
Building Modeling

Yet do much less, so much less, Someone says,
(I know his name, no matter)—so much less!
Well, less is more, Lucrezia.

Robert Browning

Possibly the most prohibiting challenge for indoor location-based services is the generation, modeling, and provisioning of sufficient environmental information for location-based applications. While it is relatively easy to define complex models of data, which are suitable for most indoor location-based service applications, these models tend to be complex enough to make the creation of them or performing calculations using them infeasible.

For indoor location-based services, there is no single and concise map information representation, which provides sufficient information for all services. Furthermore, there is no standard on map information representation, which is widely accepted. For the outdoor case, it is the case that a graph-based map representation contains sufficient information for navigation. All streets are represented as sequences of straight lines which can be interconnected by junctions. Hence, the complete street network is easily represented as a graph with vertices everywhere, where there is a sufficient deviation from a straight line and edges in between. Vertices with a higher degree than two (one incoming edge and one outgoing edge) are called junctions. For finding shortest paths and mapping erroneous location measurements to the street network, this information is sufficient. For indoor navigation systems, a graph representation of the walkable space is either complex using a lot of vertices or inaccurate. Furthermore, the higher the number of vertices, the better the abstraction of free space (e.g., there is a graph vertex near to any measured location). However, the higher the number of vertices, the more does a shortest path scrape along walls.

As a consequence, hybrid approaches have emerged, which use a combination of a graph representation, a set representation, and a purely geometric representation in a vector map or raster map fashion. However, these hybrid approaches can suffer from inconsistency unless they are generated from a consistent common dataset. Furthermore, the hybrid layers need transcoding techniques, which can map a location in one representation to a location in another representation. In these situations, a generalization of concepts is seldom a problem. For example, it is easy

© Springer International Publishing Switzerland 2014
M. Werner, *Indoor Location-Based Services*, DOI 10.1007/978-3-319-10699-1_4

to assign a room name to a coordinate in a map. However, it is not clear how to do the opposite. How can one assign a two-dimensional coordinate to a room that is possibly modeled as a polygon? The most natural choice, taking the center of all vertices of the room, does not even lie inside the room in some cases.

These complexities also arise from another very general fact. Depending on the type of positioning system and the interface definition of this positioning system, a location might be something completely different. Some systems return points, others return a polygon, yet others return arbitrarily shaped uncertainty regions representing probability of presence at a specific location.

This chapter discusses different map representations and their natural use in indoor location-based services. It is also of great interest how to find specific topological relations inside the map representation, and thus, some information is being provided on efficient topological and geometrical queries for map information. Finally, recent standardization efforts are explained.

4.1   Coordinate Systems

For representing a part of the world inside a map, one usually has to fix a coordinate system in which to represent the relevant part of the world. For indoor location-based services, most often orthogonal coordinate systems are sufficient as buildings are small enough to ignore earth curvature. However, the most natural type of location is provided by GPS before entering the building or even by high sensitivity GNSS inside the building. GPS and other large-scale systems use a spherical coordinate system to represent coordinates all over the world (except at the poles) with sufficient accuracy. Hence, with the choice of a coordinate system for the indoor location-based service area, one should also fix a projection from local to global coordinates and vice versa.

With respect to indoor location-based services, two general classes of coordinate systems are to be distinguished. Firstly, geometric coordinate systems are able to model a given space completely by assigning coordinate vectors to locations. Secondly, symbolic coordinate systems are used to have a discrete set of locations. These sets of locations often recover semantics of buildings such as rooms or floors.

4.1.1   Geometric Coordinate Systems

Two choices of geometric coordinate systems are particularly important: firstly, orthogonal coordinate systems, in which locations are represented as coordinate vectors with reference to a set of orthogonal vectors, and secondly, spherical or polar coordinate systems, in which locations are represented with respect to the center of a sphere and are typically given by a set of angles and a distance to the center of the sphere or circle.

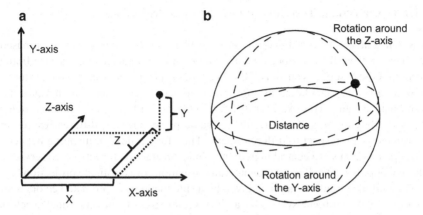

Fig. 4.1 Coordinate systems. (**a**) A Cartesian coordinate system. (**b**) A spherical coordinate system

4.1.1.1 Cartesian Coordinate System

Let $V = \mathbb{R}^n$ be a n-dimensional vector space over the field of real numbers. A Cartesian coordinate system is given by the selection of a basis for V and an ordering of the elements of the basis $b_1, b_2, b_3, \ldots$. Every element from $v \in V$ possesses a unique representation as a sum of the elements of the basis

$$v = \sum_{i=1}^{n} \alpha_i b_i,$$

and the ordered sequence (α_i) are called coordinates of v. This construction is often performed with respect to a pairwise orthogonal basis of vectors of length 1. Figure 4.1a depicts typical Cartesian coordinate systems. This choice of coordinate system has some very useful consequences for calculations with coordinates: many geometric operations such as translation or distance calculation translate to very simple expressions for the involved coordinate vectors.

4.1.1.2 Spherical Coordinate System

A spherical coordinate system is usually defined by fixing a center point $c \in V$ and different planes in which angles are measured. Every location is then specified by giving the Euclidean distance to the chosen center point, which is often called height or elevation and several angles. A very simple and illustrative case is given by the two-dimensional polar coordinate system. Figure 4.1b depicts such a coordinate system. Each point in the plane can be specified by the distance and angle with respect to the fixed axis drawn in the picture. This can be extended to higher dimensions. Figure 4.1b depicts a spherical coordinate system for three-dimensional space as it is often used in mapping locations on the world.

4.1.1.3   Coordinate Transformations and Map Projections

A single choice of coordinate system is sufficient for outdoor location-based services, and WGS 84 has been widely adopted as the underlying reference frame of major GNSS services such as GPS. For indoor navigation, Cartesian coordinate systems are a better choice as most map information is modeled in Cartesian coordinate systems. However, for different buildings, different coordinate systems are typically chosen, and as a result, the same place can have different Cartesian coordinates with respect to different maps. Therefore, a flexible mapping between coordinate systems is a central ingredient of map abstraction for integrated services. These mappings are called coordinate transformations or coordinate projections.

A coordinate transformation is given by a rule mapping coordinate vectors with respect to one coordinate system for a given vector space V into coordinate vectors for the same location in the same vector space with respect to a different coordinate system. A simple coordinate transformation between the polar coordinate system in Fig. 4.1b and a Cartesian coordinate system is given by the map

$$(x, y) \rightarrow (\sqrt{x^2 + y^2}, \arctan(y/x))$$

with the inverse mapping

$$(d, \phi) \rightarrow (d \cos(\phi), d \sin(\phi)).$$

Coordinate transformations can easily become complicated, when the reference systems are nontrivial. Map projections used in real-world map data are standardized, and for these standard coordinate systems, sophisticated software libraries are available, which handle coordinate transformation issues. One of the most important projects is given by the *PROJ.4 – Cartographic Projections Library* [6].

Sometimes, mappings similar to coordinate transformations are being used, which are not bijective and, hence, lose some information. A very common example is given by removing the elevation variable from GPS readings. In most cases, the height is actually not being used in navigation applications. Mappings of this type are called projections, as they usually satisfy the property that they are idempotent: applying a projection twice does not change the result. In a formal sense, a projection is a linear map $P : V \rightarrow V$ such that $P^2 = P$. Furthermore, projections are often designed such that their image is contained in a proper subvector space. For the example of forgetting the height information, the projection P can be given as a matrix as follows:

$$P = \begin{pmatrix} 1 & 0 & 0 \\ 0 & 1 & 0 \\ 0 & 0 & 0 \end{pmatrix}.$$

In this case, the last coordinate is always mapped to zero, and hence, the image of the projection is contained in the subvector space represented by the first two coordinate vector entries. As another example, Fig. 4.1b depicts two different, equivalent coordinate systems for representing location on earth. Every location on earth can be described either by giving the three coordinates along the X-, Y- and Z-axis in a meter scale. Alternatively, every location can be described by two angles latitude and longitude and the elevation above the reference ellipsoids surface. It is clear that each coordinate system can represent each point uniquely, and the coordinate transformation is a map mapping a vector containing the X-, Y-, and Z-coordinates to a vector containing the angles and elevation values (θ, ϕ, h).

4.1.2   Symbolic Coordinate Systems

Inside buildings, locations are sometimes not described in coordinates. This is due to the fact that there is no positioning system, which is accurate enough, and that semantic names to places are commonly used in communication between humans. The concept of naming spaces with their semantic use (e.g., kitchen is a place to cook) leads to a notion of location well suited for human–computer interfaces and compatible with coarse positioning systems which do not provide sufficient quality if used with geometric coordinates.

Symbolic coordinates are usually fixed by defining a finite set of labels and a mapping between location, and labels. Sometimes, labels are used only to describe the semantics of a location and sometimes they are chosen fine enough to differentiate between distant places with equal semantics.

Figure 4.2 depicts a very simple floorplan and assigns labels A to P to different semantic locations. These locations include typical rooms and, special rooms such as rooms with two windows or even without windows, toilets, and staircases. A symbolic coordinate system for this floorplan could consist of the label assignments of Table 4.1.

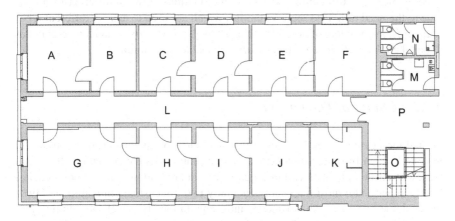

Fig. 4.2 A simple building plan with symbolic coordinates

Table 4.1 Possible label
assignment for Fig. 4.2

Label	Rooms
Manywindows	A,G
Normalroom	B,C,D,E,F,H,I,J
Nowindow	K
Toiletwomen	M
Toiletmen	N
Corridor	L, P
Staircase	O

The most important drawback of using symbolic coordinates as compared to using geometric coordinates is that they do not provide any information about their relative position. Hence, symbolic coordinate systems are often augmented by additional structures describing geometry, topology, or semantic relations between labels. As geometric coordinate systems also lack information about topology and semantics, this additional information is usually defined in general via so-called location models.

4.2 Location Models

A location model is a definition of the way in which location information is stored and retrieved. In general, a location model is based on several choices for the representation of different location information of different entities. In general, different representations are needed to provide results for different queries about location information. For example, a purely geometric location model will not be able to calculate a distance between two semantic locations given by names of streets and house numbers. On the other hand, models which are only based on street names have difficulties expressing the relations between different locations. Due to the need of hybrid representations, which are used to answer different types of queries, the problem of possible inconsistency quickly arises, and hence, care has to be taken that such inconsistencies are reduced to a minimum.

4.2.1 Choice of Dimension

For indoor location-based services, different choices of dimension are common. On the one hand, a full three-dimensional representation has the advantage that it can represent all complexities of buildings in full detail without any inconsistencies. On the other hand, most map data is actually two dimensional in nature, and two-dimensional maps are easier to understand computationally as well as cognitively by the user. However, two-dimensional maps do not suffice to represent a building with multiple floors. A compromise is often represented by a choice of 2.5D maps. These 2.5-dimensional maps usually consist of a set of two-dimensional maps, which are stacked to provide a third dimension for different floors. In this case, each location consists of a coordinate vector containing two real coordinates and an integer index coordinate. A location might then be (11.3, 6.2, 3) meaning a coordinate of 11.3 on the X-axis and, 6.2 on the Y-axis inside the two-dimensional map representing the third floor. In addition to the two-dimensional floor representation, the interconnection between different 2D slices is explicitly modeled for staircases, elevators, and the like.

These choices of dimension are also represented in positioning systems. Numerous positioning systems use different algorithms, measurements, and sensors to find the current floor inside a building, and once they have found the correct floor, they perform a two-dimensional localization inside this floor. Moreover, a lot of algorithms perform much better in two dimensions. For example, a particle filter needs considerably less particles for a sufficient representation of some probability distribution as these particles can only move in two dimensions rather than in three dimensions. Sometimes, 2.5-dimensional maps and even two-dimensional maps are annotated with extrusion information, which allows for a useful three-dimensional representation of data, though the vertical direction is limited to extrusions from the underlying two-dimensional map. Noticeably, this representation has been used by early PC games due to the simplicity of calculations, rendering, and reasoning. Ray casting-based ego shooter games such as "Doom," "Duke Nukem," and "Wolfenstein" all rely on 2.5-dimensional maps, sometimes even with additional constraints of orthogonal walls, though the user experiences a three-dimensional world. Figure 4.3 depicts examples of typical map representations.

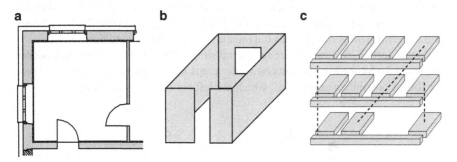

Fig. 4.3 Different maps in varying map dimensions. (**a**) A two-dimensional floorplan including symbolism for doors. (**b**) A simple 3D model of a room including a window and doors left open. (**c**) A 2.5D block model in which a room is represented by a rectangular block in 3D and interconnections are given

4.3   Vector Maps

A vector map is a representation of a map as a list of vector primitives. A vector primitive is an n-dimensional shape, which can be defined through a finite set of geometric coordinates. Typical vector maps include at least one primitive for each dimension, for example, a primitive for representing a point as a zero-dimensional object, a primitive representing a line as a one-dimensional object, a primitive representing solid triangles as a two-dimensional object, and, for three-dimensional maps, a primitive, for example, representing a solid cube. Figure 4.4 shows these primitives. However, more complicated primitives are possible and sometimes used such as spheres, ellipsoid, or parametric lines and surfaces such as NURBS and splines.

In computer graphics applications, it is common to represent every scene as a set of triangular faces for rendering. Rounded objects have to be tessellated into small triangles such that the discretization error is low enough such that the objects appear to be truly round. This process is in parallel to some vector map representations, where complex objects such as circles and arcs are well represented for software, which supports them, and contain an approximation by lines for simpler software.

4.3.1   Basic Algorithms for Vector Maps

One of the most important drawbacks of vector maps is the complexity of even simple calculations. Without additional data structures, calculating the distance between a point and the next line is possible only by calculating the distance between the query point and each line of the vector map. To reduce this scalability problem, however, spatial indices can be used to efficiently prune the list of lines and to perform most calculations only against a small set of candidate lines.

In this section, we concentrate on several simple algorithms to allow the reader to understand why some operations on vector maps are getting easily complex. There are lots of possible optimizations, which are left out here to gain clarity.

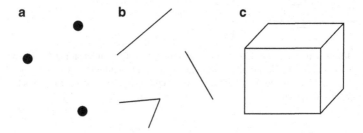

Fig. 4.4 Elementary primitives for vector maps. (**a**) Points. (**b**) Lines. (**c**) Cube

4.3.1.1   Point–Line Distance Calculation

It is a ubiquitous query to find the distance between a point and a line or a point
and a set of lines. For indoor navigation applications, this query can be used to find
the nearest edge in an embedded navigation graph to provide a mapping from space
to the edges of a navigation graph. This can especially be useful, if the navigation
graph has a lot of very long edges and the position of the mobile user inside these
edges is also of interest. Figure 4.5 depicts this situation. The location of the mobile
user is depicted as a small circle; points of interest are drawn as stars. The star
directly beneath the user is of more interest than the stars near to the segment start
point or segment end point.

We will limit the discussion to the two-dimensional case using the normed vector
space $\mathbb{R}^2$ together with the norm

$$||p||_2 = \sqrt{p_1^2 + p_2^2}.$$

This norm induces a distance given by

$$d(p,q) = ||p - q||,$$

which we will use to calculate the distance between any two points $p, q \in \mathbb{R}^2$. Let
$\mathcal{L}$ be a line segment starting at a point x and ending at a point y. Then this set of
points can be represented as the image of a function on the unit interval:

$$\mathcal{L} : \phi(\lambda) = x + \lambda u.$$

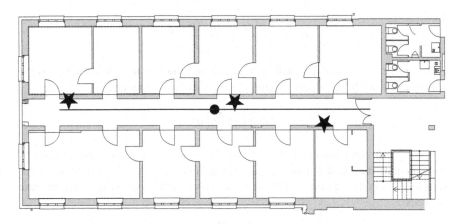

Fig. 4.5 An in-segment point of interest. The *circle* denotes the mobile device, and the *stars*
denote points of interest

The direction vector u is given by the vector from the start point of the segment pointing to the end of the segment, e.g.,

$$u = y - x.$$

The parameter λ can be seen as a time for traveling along the line segment. At time instant $\lambda = 0$, one is located at the starting point x; at time instant $\lambda = 1$, one finds himself at the ending point y.

For calculating the distance between a point p and this line, one is left with an optimization problem as follows.

Find the parameter λ for which the distance $d(\phi(\lambda), p)$ is minimal. Using the fact that the square root function is strictly monotonous, this minimum for the expression $d(\phi(\lambda), p)$ is achieved for the minimum of the argument of the square root inside the norm, which can easily be expressed as a scalar product as follows:

$$||\phi(\lambda) - p|| \to \min \Leftrightarrow (\phi(\lambda) - p)^T (\phi(\lambda) - p) \to \min.$$

Setting $z = x - p$, we are left with minimizing

$$z^T z + 2\lambda u^T z + \lambda^2 u^T u.$$

The first derivative with respect to λ is given by

$$\frac{\partial}{\partial \lambda} \left(z^T z + 2\lambda u^T z + \lambda^2 u^T u = 2\lambda u^T u + 2u^T z \right),$$

which has a single root

$$\lambda^* = -\frac{u^T z}{u^T u} = -\frac{u^T (x - p)}{u^T u}.$$

At this point, the function actually achieves a minimum as the second derivative is simply given by $2u^T u$, which is a sum of squares of real numbers:

$$2u^T u = 2 \sum u_i^2.$$

For the calculation of the distance between p and $\mathcal{L}$, we can use this discussion and calculate the nearest point to p on the infinite line given by extending $\mathcal{L}$ in both directions. The distance calculation can now be completed with a case-by-case analysis:

- $0 \leq \lambda \leq 1$: The distance is given by $||\phi(\lambda^*) - p||$.
- $\lambda < 0$: The nearest point on the line segment is the starting point, and the distance is given by $||\phi(0) - p||$.

- $\lambda > 1$: The nearest point on the line segment is the ending point, and the distance is given by $||\phi(1) - p||$.

This example shows that even simple calculations with vector maps can lead to complicated expressions including several case distinctions.

4.3.1.2 Line–Line Distance Calculation

Now that we are able to calculate the distance between a point and a line, we can try to extend towards calculating the distance between two lines. This can be achieved very similar to the case of point–line distance by varying the reference point p along the second line. Let two lines $\mathscr{L}_{1,2}$ be given as follows:

$$\mathscr{L}_1 : \phi(\lambda) = x + \lambda u$$
$$\mathscr{L}_2 : \psi(\tau) = p + \tau v.$$

Then we are left with minimizing

$$d(\mathscr{L}_1, \mathscr{L}_2) = \min_{\lambda, \tau} ||\phi(\lambda) - \psi(\tau)||.$$

This can be reduced to a minimization with respect to a single parameter using the result from the previous section. Let therefore $\Lambda(\tau)$ be the parameter of the nearest point on $\mathscr{L}_1$ with respect to $\psi(\tau)$:

$$\Lambda(\tau) = -\frac{u^T (x - \psi(\tau))}{u^T u}.$$

Then we are left with minimizing

$$d(\mathscr{L}_1, \mathscr{L}_2) = ||\phi(\Lambda(\tau)) - \psi(\tau)||,$$

which only depends on τ. This is also possible by finding an optimal τ. This can be computed as the root of the first derivative of the scalar product of the inner of the norm, e.g.,

$$\frac{\partial}{\partial \tau} (\phi(\Lambda(\tau)) - \psi(\tau))^T (\phi(\Lambda(\tau)) - \psi(\tau)) \to \min.$$

This provides an optimal τ and, an optimal $\lambda = \Lambda(\tau)$, hence two points $\phi(\lambda)$ and $\psi(\tau)$. However, a complicated case-by-case analysis is needed for cases where either of λ or τ is pointing outside the respective line segment.

4.3.1.3 Point in Polygon

For navigation applications, it is often very important to map raw location data given as geometric coordinates into more meaningful representations of location. For example, rooms can often be modeled as polygons, and finding the room given some coordinates is a common problem. Geometrically this problem boils down to efficient algorithms testing whether a point is inside a polygon. Fortunately, there is a quite simple algorithm. However, this algorithm is based on a very deep insight into two-dimensional geometry called Jordan curve theorem:

Theorem 4.1 *Let C be the image of a closed, non-selfintersecting, continuous loop in R^2 given as a function $\phi : [0, 1] \rightarrow R^2$ with $\phi(0) = \phi(1)$. Then this curve C splits $R^2 \setminus C$ into two connected components, one of them bounded, and the other unbounded. Furthermore, C is the boundary of both sets.*

This essentially says that for each loop without self-intersection, especially for each polygon without self-intersection, each point is either inside the polygon or outside. Moreover, this gives a simple algorithm to find for a given point and a polygon, whether the point is inside the polygon or outside: casting an infinite ray from the point p into any direction and intersecting this ray with any edge of the polygon leads to one of three cases:

- There is an even number of intersections.
- There is an odd number of intersections.
- There are infinitely many intersections.

In the first case, the point is clearly located outside the polygon, in the second case inside. In the third case, the test has to be repeated with a different angle. Figure 4.6 depicts these cases.

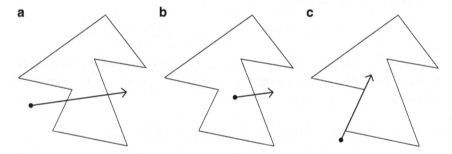

Fig. 4.6 Different situations for a point in polygon test based on Jordan theorem. (**a**) Case 1: even number of intersections. (**b**) Case 2: odd number of intersections. (**c**) Case 3: infinite number of intersections

4.3.1.4   Further Readings

The previous section has given some very basic algorithms common to processing maps in geometric coordinate systems. A lot of books cover these algorithms in detail stemming from different areas: the basic algorithms can be found in most books on computational geometry, such as [2]. The area of moving object databases also covers topics of query processing and indexing for speeding up queries against databases with lots of elements [7].

4.3.2   Raster Maps

Vector maps can be used to represent the world with arbitrary precision at the cost of complexity and difficult error treatment. In raster maps, the map information is not represented by primitives but using a discretized grid of integer pixels. Raster maps can directly provide a trade-off between accuracy of representation and size by choosing appropriate pixel sizes. Usually, in so-called occupancy grid applications, the raster map is considered as a matrix containing zero entries representing an empty cell and different colors for different nonempty cells. In this way, the location of obstacles such as walls and furniture is easily modeled in two dimensions. It is worth noting that three-dimensional raster maps are still impracticable for their memory footprint for most applications and that full three-dimensional modeling usually prescribes the use of vector maps.

Raster maps are limited not only with respect to dimension and accuracy. It is also difficult to model map data that is not axis parallel. As depicted in Fig. 4.7a, a straight line with a width of one pixel can easily be modeled horizontally and vertically. For a diagonal line, however, there are spots, where white pixels on both

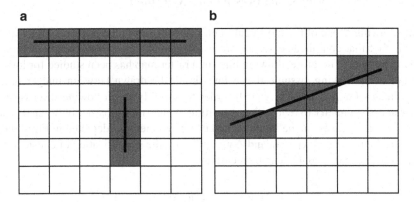

Fig. 4.7 Lines in raster maps. (**a**) *Horizontal* and *vertical lines* in raster maps. (**b**) A *diagonal line* without anti-aliasing

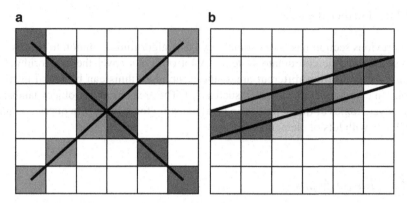

Fig. 4.8 Intersection artifacts and anti-aliasing. (**a**) Two intersecting lines whose pixel sets do not intersect. (**b**) A line drawn using anti-aliasing

sides of the line are adjacent as depicted in Fig. 4.7b. This can lead to situations as depicted in Fig. 4.8a, where the pixel sets of two intersecting lines actually do not intersect. In order to overcome these problems, it is common to use anti-aliasing approaches in which the color of a pixel next to a line is inferred from the percentage of occupancy of the pixel by the line blending between the pixel color and the line color. For humans, the perception of raster graphics generated this way is almost perfect, as the line seems to have the correct extent. However, anti-aliasing induces additional complexity to raster image algorithms.

In order to illustrate some basic operations on raster maps, the following section explains some of the algorithms for which a raster map can be better suited than a vector map. These include line drawing and filling connected areas.

4.3.2.1   Line Drawing Using Bresenham's Algorithm

Line drawing is one of the most central elements of computer graphics. This is due to the fact that complicated scenes are usually constructed from complex objects bound by lines. Therefore, drawing lines in a raster map has been studied for long.

For the following, assume that a line should be drawn between integer pixel coordinates (x_0, y_0) and (x_1, y_1). In order to simplify the following discussions, we can use symmetry to reduce the exposition of algorithms to lines between $0°$ and $45°$. The other cases can be implemented using symmetry as depicted in Fig. 4.9a.

Introducing $\Delta x = x_1 - x_0$ and $\Delta y = y_1 - y_0$, the simplification of considering angles between $0°$ and $45°$ can be recast as

$$s = \frac{\Delta y}{\Delta x} \in [0, 1].$$

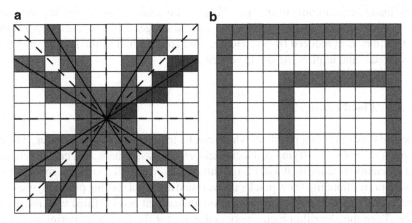

Fig. 4.9 Line painting and flood fill algorithm. (**a**) Line painting for lines from 0° to 45° suffices due to symmetry. (**b**) Flood fill movement inside a raster map

A very simple approach is given by calculating this vertical slope s of the line with respect to one step into the X direction. Then, the algorithm can increment the X value and add this slope to a Y variable until the line ends.

This simple algorithm, however, needs floating point computation in order to track the Y value during line drawing. Furthermore, the computation of s will contain rounding errors leading to the case that the line might not hit the intended endpoint. For faster implementation, the Bresenham algorithm can be used, which only needs addition of integer numbers and no multiplications or divisions inside the drawing loop. This leads to very fast implementations. As a side effect, it minimizes rounding errors, which can lead to problems in the naïve formulation above using the quotient $\Delta y / \Delta x$.

The algorithm works as follows: into one direction (the so-called fast direction or axis-parallel direction), a unit step is performed. From time to time, a step into the other direction (the so-called slow direction or diagonal direction) has to be performed. These steps are controlled by representing the line in the following form and tracking the actual value for the approximated line on the left hand as an error variable E.

$$0 \overset{!}{=} E = \Delta x(y - y_0) - \Delta y(x - x_0)$$

Going a step into the X direction decreases the left-hand side by Δy. Once the left-hand side falls below zero, a diagonal step is introduced and the error variable is increased by Δx. For lines in the correct octant (0–45°), we know that $\Delta x \geq \Delta y$. Hence, a diagonal step increases the error above zero, and consequently, a step into the X direction can follow. To complete the algorithm, a suitable initial value for E must be defined. A simple choice of initialization can be motivated from a special case: if $\Delta y = 1$, the line contains a single vertical step. This could be expected to be made in the middle of the line. This can be achieved by initializing the error variable with $\Delta x/2$, possibly rounded to the next integer value: for each step into the X direction, E is reduced by one. Hence, after roughly half of the line, $E = \Delta x/2$ crosses zero and implies the single vertical step.

4.3.2.2 Flood Filling

Another very important problem, which can be tedious to solve for arbitrary vector maps, is given by classifying connected spaces. For raster maps, this can effectively and efficiently be done by flood filling. This operation can fill connected space with a new color by "flooding" new pixel values around an initial starting point. Figure 4.9b depicts this situation. For every invocation, the current pixel is filled with the given color. Then the algorithm recursively continues for the four neighboring pixels as depicted in the figure. If these have the same color as the initial pixel had, the pixel is filled and the recursion is repeated. If the color is different, an obstacle has been found. The pixel is not filled with the new color and the recursion ends.

This algorithm fills all pixels reachable via horizontal and vertical movements inside the same color as the start pixel. For location-based services, it can be reasonable not to actually paint into the map but to enumerate the connected space in order to check, whether two locations are inside the same connected component. In this case, a bitmap of flags can be used in order to track locations, which have already been seen by the algorithm.

In this simple form, the recursion can get very deep and even impracticable. Therefore, some heuristics can be used to iteratively fill the surroundings of some pixel, e.g., by rotating clockwise. Only, when this iteration ends, the recursive part is started.

4.4 Environmental Models

The previous sections discussed the two main approaches to modeling indoor space geometry. Vector graphics come with great flexibility and expressiveness. Computing with vector maps can, however, become complicated. Raster maps are practically limited to modeling two-dimensional spaces. However, they can be useful for their inherent ability to deal with small errors and their simplicity in calculations.

Both approaches, however, do not suffice to provide a sufficient data basis for indoor location-based services, as they are limited to geometry. Other aspects such as functionality or topological relations are only implicitly contained through symbolism, which is hard to exploit in computer programs. Therefore, the geometric representation is often extended and integrated with other information about a building for getting a complete environmental model supporting applications. It is, for example, obvious that raster maps cannot model buildings. Instead, a raster map per floor can be used, and the interconnections have to be modeled in a different way.

For the specification of the interconnections and topology of different sets of geometric map information, higher-order environmental models can be constructed which fall into one of the following three categories: set-based environmental models define a finite set of atomic places such as room names. Complex places such as a floor consisting of several buildings are then modeled as sets containing atomic places. Though sets provide concepts such as intersection and set similarity, they lack support for distance calculation, the neighbor relation of rooms, and distance calculation. Therefore, graph-based environmental models can be used. In order to support more applications, these two approaches, set based and graph based, can be combined to hybrid environmental models.

Before the discussion of these concrete approaches to providing environmental models, it is reasonable to discuss what an environmental model should provide. Firstly and most obviously, an environmental model should provide a geometric representation of the indoor location. This can be done using the two approaches presented in the preceding section using raster maps or vector maps. On top of this purely geometric information, there are the following queries, which are hard or impossible on geometric information and which motivate the introduction of environmental models.

Environmental models should provide support for the following operations: finding, whether two different locations are connected with each other. Is it possible to move from location X to location Y? This general relation is called *connected-to* relation. It is often true and not too useful for an application in its simplest form: every location that is sensibly used by a location-based service will be reachable from some entrance to the building. Therefore, all such locations will be connected. However, when it is raining, the query whether the location X is connected to location Y through an indoor connection can be useful for a context-aware navigation system. Instead of adding this complication to the *connected-to* relation, it is customary to add a *range query*. The range query returns an environmental model limited to a range. For the example, this range query would be performed with the range of all indoor places, and then the *connected-to* relation would exactly express the indoor connectedness of places. Another topological relation is given by neighborhoods. The *neighborhood query* is defined to return the *neighborhood* of a place. This query shall provide insight into the direct connections between different places. The neighborhood query returns an environmental model of the same type and level of detail, which contains the model used for the query as well as all neighbors. In this sense, the neighborhood of a building consists of the building itself and the buildings located next to the buildings, while the neighborhood of a room consists of the room itself and the rooms next to the room. Another piece of environmental information, which is often needed, is given by the *containment* of places. Does a specific room belong to a specific building? Are two buildings in the same city? The *containment* relation can be augmented by a *generalization* query. Given some environmental submodel, this operation is defined to find the smallest environmental model that contains the given environmental model. As an example, one might be interested in finding out the address and geospatial location of the building given a specific room returned by a symbolic positioning system. A common refinement of the *connected-to* relation is given by the *distance-to* relation, which returns the length of the shortest connection between two places. This is a common relation in navigation systems and leads to a query called *shortest path*, which returns the shortest path between two places inside an environmental model. In combination with the *range query*, this can be used to find sensible connections for specific user groups. A wheelchair user might decide to remove all staircases from the environmental model. Then the shortest path will lead the wheelchair over building objects, which he can actually use. Finally, the *distance-to* relation can be combined with the Neighborhood relation to a *nearest neighbor* query.

In summary, a good environmental model should be able to support:

- *Range queries and sub-models*
- *Containment and generalization*
- *Connected-to and neighborhoods*
- *Distance-to and shortest-paths*
- *Nearest neighbors*

4.4.1 Set-Based Environmental Models

A set-based environmental model uses sets to represent spaces and subspaces. This is a quite natural choice in many situations. For example, a city is composed of buildings, which are composed of floors, which are split into rooms. These environmental models are thus based on a symbolic coordinate system. In set-based environmental models, the smallest piece of location is modeled as an *atomic place*. Often, text labels such as room names represent these atomic places. Larger concepts emerge from atomic places as the union of sets and the concepts of subsets, intersection, and the element relations can be used to reason about these models.

A difficult design decision in deploying set-based environmental models is given by the choice of the size and properties of atomic places. In the case of using set-based environmental models together with some geometric information, it is reasonable to label, for example, all bounded and connected areas with a different label and build a more semantic layer by capturing different rooms as unions of different bounded and connected areas. Though this approach seems to be unproblematic, one can easily run into the following two problems: firstly, the size of the sets can become quite large, and a lot of labels have been defined, which are never used in for location-based services. Secondly, the geometrical basement leads to situations, where some atomic places are part of different semantic places. This can quickly lead to inconsistencies, when the intersection of two rooms shall cover generalization concepts such that a room is part of a floor.

Figure 4.10 depicts a set-based environmental model in which the atomic places are given by rooms, and labels represent these atomic places. From this model, two

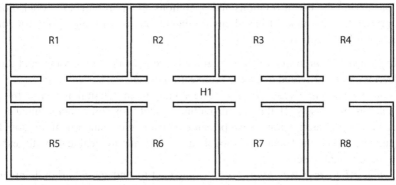

Floor1 = {R1,R2,R3,R4,R5,R6,R7,R8,H1}
G1 = {R1,R2,R3,R4,R5,H1}
G2 = {R5,R6,R7,R8,H1}

Fig. 4.10 A set-based environmental model modeling $Floor1$ and two submodels modeling user groups $G1$ and $G2$

sets are derived called G1 and G2 representing user groups of the building. The inherent advantage of this set-based approach is given by the conclusion that room R5 is used by both groups. The intersection of both sets G1 and G2 contains R5 meaning that room R5 is used by both groups. This information can hardly be expressed inside a geometric map without special symbolism or standardizations. In this way, the example again shows that modeling the geometry of a building does not suffice for smart indoor location-based services.

In summary, a set-based environmental model can be used to model the concept of overlapping places given by the intersection of sets and the containment relation of places given by the subset relation. The element relation easily captures the relation between a user's actual location and the environmental model: the user is in all places in which the atomic place, which he actually is inside, is an element of. Note that it is customary to assume that a specific location of a mobile user is contained in only one atomic place in order to overcome problems of uncertainty.

Unfortunately, set-based environmental models do not suffice for location-based services as they cannot cover aspects such as distances between rooms: if several shared meeting rooms are available, it is a common location-based service to find out a reasonable one, which is near to all users attending the meeting. It is, however, impossible to model this concept using set-based environmental models only. Therefore, graph-based models have been proposed, which are also based on a finite set of atomic places. However, these places are connected with each other, where a direct connection is possible and annotated with the distance or cost of taking this connection. Therefore, graph-based models can be used to assess the distance and solve a lot of optimization problems in the context of location-based services.

Considering the list of relations and operations an environmental model should support, we find out that set-based environmental models are quite good for some aspects, but lack support for others:

- *Range Queries and Submodels:* Submodels are provided by subsets, and range queries can be provided using the intersection of sets.
- *Containment and Generalization*: The containment relation is given by the containment of sets. If the sets providing models for some hierarchical level are known, generalization is also possible by using the containment relation: the generalization is the "smallest" model on a higher hierarchical level with respect to containment of sets.
- *Connected-To and Neighborhoods:* Connected-to and neighborhoods are not available. However, they can be emulated by modeling neighborhoods manually as generalizations of atomic places. However, this is only an implicit form of a graph-based environmental model to be described in the next section.
- *Distance-To and Shortest Paths:* Distance cannot be captured in set-based environmental models, and therefore, shortest paths are not defined.
- *Nearest Neighbors:* Similarly, as the concept of distance is lacking, nearest neighbors are limited to manually modeled neighborhoods.

In summary, set-based environmental model are useful for roughly half of the requirements. Especially, the simplicity of sets makes range queries and submodels very convenient and leads to simple containment and generalization.

4.4.2   Graph-Based Environmental Models

A graph-based environmental model is constructed by giving a definition of what a vertex is and a rule of when two vertices are to be connected. Here, a graph G consists of a set $V(G)$ of vertices and a subset $E \subseteq V \times V$ of edges. It is useful to imagine a vertex as a point and edges as arrows pointing from the source point to the destination point.

Given a building floorplan, there are plenty of different possibilities of designing a navigation graph for the building. The choice of the actual type of graph depends on the amount of information available as well as on the intended use case. From an environmental model perspective, the simplest case is given by a navigation graph, which is modeled by hand. Vertices are added to the building floorplan, and edges are used to interconnect them reasonably. In this scenario, it is possible to draw beautiful graphs whose edges lie inside the free space and can readily be used for visualization. However, modeling the navigation graph by hand is a complicated task, and every change to the building has to be manually integrated into the graph. Therefore, automated systems are needed, which can generate reasonable navigation graphs from available building information such as floorplans.

If the building floorplan is given as a set of polygons, for example, and if the space without drawings is actually walkable space, then a simple idea is given by connecting every pair of vertices of the polygons as long as the edge is completely contained in walkable space. This definition leads to the so-called corner graph. The main strength of the corner graph is its simple construction and its completeness. For polygonal maps, the corner graph always provides the shortest paths, when the source and target location are mapped to polygon vertices.

Figure 4.11 depicts an example floorplan and the beginning of a corner graph. In this figure, only those edges of the corner graph are depicted which start inside the room. It is obvious that a corner graph can become quite complicated.

In many cases, not only room polygon information is available. One type of information, which is often explicitly modeled, is the location of portals. Portals are used to connect rooms with each other and can also be used to model vertical passages between different floors. Examples of portals include doors, staircases, elevators, and escalators. In the case that portal information is available, these portals can be used to provide a starting point for graph generation, as portals are usually those elements of building connecting different places.

In Fig. 4.12, the doors are modeled as portals, and the graph contains one edge modeling, each such portal depicted slightly bolder than the other edges. In this case, the portal is directly connected to all line-of-sight vertices. In order to model rooms, one vertex per room has been added to the graph. In this simplicity, this works only if the rooms are convex or edges are allowed to cross walls. In non-convex buildings, the approach is usually coupled with another approach modeling the interior of each room. For example, a recursive corner graph construction starting at the portal vertex inside the given room can be used to model arbitrary interior spaces. Modeling portals is usually performed in a manual way and leads to a medium modeling effort. The resulting graph, however, enables several use cases: as each shortest path contains all portals being used, it is quite easy to detect changes of semantic places and, hence, to understand the relation between locations. Even when a very complicated polygon leads to hundreds of edges being traversed inside a room, the

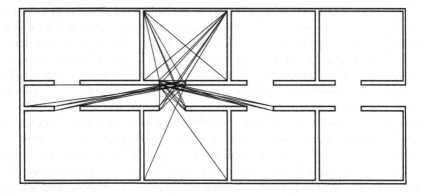

Fig. 4.11 A partly completed corner graph

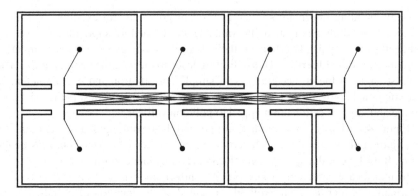

Fig. 4.12 A portal graph covering door portals

essential information of changing rooms is available from the graph. Parallel to starting navigation graph construction with portals, it can be reasonable to start with vertices defining rooms.

Figure 4.13 depicts this case: again, modeling is performed manually in most cases. The resulting graph, however, contains only rooms and edges between rooms and can, hence, be used to generate sequences of rooms to enter. Note that the edges do not need to lie inside free space in this case. However, bootstrapping a graph from portals and room vertices leads to hybrid models as depicted in Fig. 4.14.

The effort in generating such graphs is high enough that one can reasonably expect the rest of the graph being drawn by hand, too. In these cases, it can be possible to augment the graph with additional information about the nature of an edge: especially, information about landmarks passed by the edge can be useful for instruction generation.

While the corner graph assumes the building being given by means of polygons and the room interconnection graphs assume knowledge of semantic places and points of interests (e.g., doors and rooms), more general approaches are needed for cases, where the floorplan is given as a raster map or where the set of points of interests is not known at the time of modeling the building or changing quickly. For this situation, it is reasonable to assume that navigation graphs are constructed from an occupancy grid, which can easily be generated from any type of environmental model. This occupancy grid is a raster graphic containing white pixels for free space and black pixels for occupied space. It is now possible to use a grid construction for generating a navigation graph. A grid of points is laid over the environmental model, and two neighboring grid points are connected if and only if the edge in between is contained in walkable space. Using the eight-corner system of orientation including up, down, left, and right as well as their diagonals, one can generate navigation graphs such as the graph depicted in Fig. 4.15. These graphs fill out the complete walkable space with vertices and edges in a given grid resolution. This leads to the possibility of navigating between all pairs of coordinates by rounding the coordinates to the nearest grid point, which is reachable via walkable space.

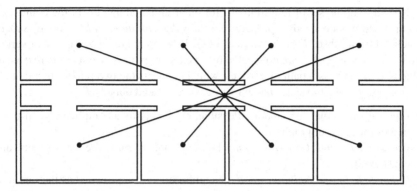

Fig. 4.13 A room graph with edges not contained in walkable space

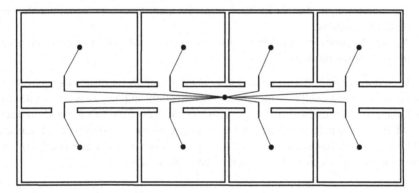

Fig. 4.14 A hybrid graph modeling rooms and door portals

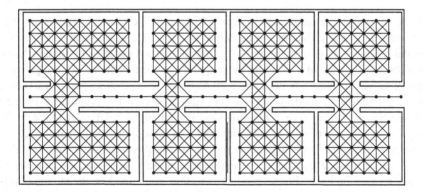

Fig. 4.15 A regular grid graph based on the eight-corner system

Considering the set of requirements for environmental models in the context of graph-based models needs a preparation: the *subgraph induced by a vertex subset* is defined for a subset T of the vertices of G. It is defined to be a graph with vertex set T. Two vertices of T are connected by an edge in two cases: either, there was an edge in G or there is a path of vertices, which connects the two points such that all vertices expect that the start and end vertices are located outside T.

- *Range Queries and sub models:* Possible, but involving a complicated construction of induced subgraphs.
- *Containment and Generalization*: Difficult unless only the vertex sets are considered.
- *Connected-To and Neighborhoods:* Neighborhoods are provided as the one-hop neighbors in the graph, that is, the end vertices of outgoing edges of a vertex. The connected-to relation is given by a sequence of neighborhoods.
- *Distance-To and Shortest Paths:* The edges of a graph connect places and can be assigned a weight. In this situation, efficient algorithms for calculating shortest paths are available.
- *Nearest Neighbors:* A breadth-first search can be used to find nearest neighbors in graph-based models.

In summary, graph-based models have natural support for distances, navigation, and connectedness, while containment, generalization, and range queries need some complicated constructions. In this sense, graph-based environmental models and set-based environmental models provide natural support for different types of queries. Moreover, graph-based environmental models can be assigned an underlying set-based environmental model by forgetting edge information.

4.4.3 Hybrid Approaches

While, for some queries related to indoor location-based services, graphs are very efficient, they have difficulties in modeling free space or hierarchical concepts. Set models allow for easy definition of hierarchical structures such as buildings, floors, and rooms. They are also good on range queries as long as the ranges are given as sets of atomic locations themselves. Range queries with arbitrary geometric ranges are impossible. Set models are—in general—limited to the granularity of their atomic places. Furthermore, the complexity of calculations and the size of the model depend on this granularity. Combining both to a hybrid environmental model can efficiently support all topological queries. However, this introduces the possibility for severe inconsistencies between the different models. Moreover, the coupling to geometry is quite loose. For set models as well as graph models, the atomic places for sets and vertices for graphs can be bound to a geometric location. Unfortunately, the type of location is difficult to choose and depends on the semantics of atomic places. While it could be nice to represent rooms by polygons, some labels might be better represented by points, circles, or other geometric objects. Hybrid location models should therefore provide a flexible linking to geometric models or sub models. This introduces a new problem of managing the relations between different models added to a hybrid location model in order to support a specific query or feature. This problem is given by translating between geometric and topological models. Assigning a geometric location to a topological entity is called *geocoding*. Assigning a topological location to a geometric entity is called *reverse geo coding*.

Topological information is often used coarser than geometric information. While it is geometrically relevant whether a user stands at one end of a room or at the other end, the topological situation is the same for most queries: a shortest path on a room interconnection graph only depends on the location of the doors. This motivates hybrid geo-topological models based on modeling each atomic place in a geometric submodel. Modeling the large-scale effects in set-based or graph-based environmental models and providing geometrical information only locally for atomic places provide several advantages: for a mobile system, geometric information need not be retrieved for a complete building, while planning tasks such as navigation can still be performed globally. This reduces bandwidth for mobile devices dramatically. Furthermore, the actual geometric representation type (e.g., vector map, raster map) as well as the level of detail (e.g., how many primitives, which resolution) can be specified by the mobile device to best support the actual use case. This can also heavily reduce computational overhead on the mobile device. However, this approach also introduces a drawback of increased redundancy and possible inconsistency: the walls distinguishing between neighboring rooms might be modeled twice, once for each of the neighboring rooms in the room's local geometric model. If the model is very coarse, the walls could intersect or be doubled when presented to the user. When, however, the problems of inconsistency and redundancy are accepted, the complete model can be made more flexible by allowing several instances of different types of environmental information inside the same environmental model together with transcoding mechanisms between any of them. A fully *hybrid environmental model* is given by a collection of environmental models called *Environmental Building Blocks*. This can be seen as a loose coupling of several environmental models. Each model contains either geometry, set-based models, or graphs together with operations transforming an entity, location, or object in each building block into a representation in another building block.

Collecting several different environmental models inside one large meta-model provides several perspectives on the same environment. Figure 4.16 depicts an example. The example consists of three building blocks tailored to the three different aspects *visualization*, *walkable space* and *navigation*. The building block *visualization* contains a detailed drawing full of symbolism including symbols for doors and windows. This drawing is used mainly to present the environment to the mobile user. The second building block is given by an occupancy grid representation of the walkable space. It consists of black walls and white pixels, wherever a user is able to walk. The third building block contains a navigation graph connecting middle points of rooms with each other. Furthermore, different navigation graph for different user groups can be added to the hybrid model. An application then chooses

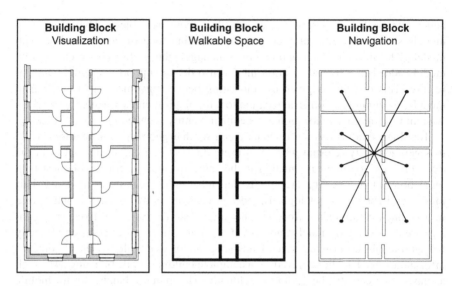

Fig. 4.16 A hybrid environmental model with three building blocks: *visualization*, *walkable space*, and *navigation*

the correct building blocks for a given purpose, and only those building blocks are used by an application, which are really needed. This opens environmental modeling to the freedom of flexibly providing environmental information to the mobile application depending on the use case, while being able to translate between different representations in cases, where the perspective has to be changed.

4.5 Geometric Nearest Neighbors and Range Queries

Hybrid models are able to support all requirements of environmental models out of the box. Set-based models can be used for range queries and containment with atomic places; graphs can be used to calculate distances and topological relationships. However, range queries as well as next neighbor queries are supported only for atomic places in set-based models or for vertices in graph models. It is, however, not needed to provide additional model types for providing full geometric nearest neighbor and range queries, as there are efficient data structures for point sets and arbitrary geometry. This section explains one such data structure and limits the discussion to points. With these algorithms and hybrid location models from the previous section, all requirements for environmental models as discussed are fulfilled.

In general, k-next neighbor search is given by efficiently returning the k nearest neighbors of some query object. For simplicity, we consider the problem of finding the nearest k points inside Euclidean space $\mathbb{R}^n$.

In order to efficiently find the nearest k points in a database given a query vector $x \in \mathbb{R}^n$, the baseline approach is given by linear search: for each vector y in the database, calculate the distance $d(x, y) = ||x - y||$ and remember the vectors associated to the k smallest distances.

However, this approach does not scale very well. For large databases, every vectors needs to be examined. Therefore, data structures have been designed with which it is possible to find small candidate sets of vectors of the database which have a chance of being along the k nearest neighbors for arbitrary query vectors. In general, linear search is then used inside this candidate set. There is a vast body of research on optimal data structures for a vast amount of different special situations. For this book, we only give one example of a data structure to show the principles with which nearest neighbor queries can be performed efficiently.

A classical data structure for this task is given by the kD-tree. It is based on recursively splitting the space using axis-parallel hyperplanes. This recursive splitting induces a binary tree structure in which each child represents the subset of all points lying on one specific side of the hyperplane. Inner nodes contain elements that are located exactly on the splitting hyperplane.

Figure 4.17a depicts this situation. The points of interest are drawn as dots. The given tree first splits in left and right using the depicted line and then further splits into up and down. In this case, the inner nodes contain the points that lie on the splitting line. This correspondence is indicated using dashed lines. All other points are associated to the leaf node, in whose spatial region they are located. Note that empty branches are possible; in the figure, the branch "up-right" is actually empty.

This tree structure can be used to speed up spatial range queries of different types by generalizing point sets to their spatial regions given by the tree splittings. For retrieving a small superset of the points lying inside some spatial region, we can take the union of all regions with which the query region intersects. But before we can exploit this fact for nearest neighbor search, we need to generate a suitable query region given a single query point.

Therefore, we first locate the last nonempty region in which the query point resides starting from the root of the tree. This leaves us with at least one candidate point. This candidate provides us with a candidate circle in which we have to look

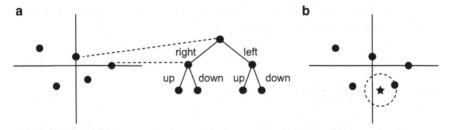

Fig. 4.17 A KD-Tree for points and its application to nearest neighbor search. (**a**) A set of points and its kD-tree. (**b**) A query point together with a candidate defines a *circle*, which has to be checked

for elements closer to the query point. Figure 4.17b depicts this situation: the point inside the "down-right" section defines a circle. It would clearly be the nearest point, if the circle did not intersect the region "down-left." This region must be explored, before we know the nearest neighbors. However, all points above the horizontal line can be left out of consideration.

The algorithm now descends into all nodes which represent spatial regions intersecting the current query circle. After that approach, the nearest neighbor has been found. Still, some other points might have been found such as the first candidate point. These points l which have automatically been found are the l nearest neighbors of the point. If $l < k$, we have to further generalize spatial regions towards the root node until we either find k points or run out of points.

The remaining part of using the kD-tree data structure for finding nearest neighbors is given by an algorithm to construct the tree. This is easily formulated recursively, that is, given a set of points S, we find a single hyperplane and split the set into sets S_l and S_r and recursively continue for each of these sets.

For kD-trees, the splitting hyperplanes are defined as follows: first, the direction of the split is calculated, and then, the split location is calculated. The split direction is given as the axis of maximal variance of data. Once this axis of maximal variance has been identified, a splitting hyperplane is constructed perpendicular to it. Now the median value along this axis of maximal variance is chosen as the splitting point. In this way, on both sides of the hyperplane remain the same number of points making the tree balanced. This splits into two equally sized sets along the direction of largest spread. Sometimes, it can be better to use the mean value instead of the median, as this can reduce the number of parallel choices leading to long rectangular spatial regions and successive parallel splits.

A lot of different data structures of this type have been proposed with different spatial regions and spatial splits of the point set. However, the basic principle of spatial indexing structures remains the same.

Similar to intersecting a sphere with a set of points as done inside the k-next neighbor search algorithm from the previous section, rather all spatial queries can be issued against a spatial index tree. The only difficulty is given by efficiently intersecting the spatial query object with the spatial ranges represented by nodes of the spatial index tree. If the query object is too complex, it is possible to simplify it by, for example, using the minimum enclosing rectangle or the convex hull of the query object. This, however, leads to more possible cell intersections and hence to a larger candidate set during query processing.

4.6 Standardization

As we have seen in the previous sections, it is quite difficult to provide environmental information for indoor location-based services. A lot of different perspectives and problems leads to interest in different representations of the same building.

A lot of standards have been defined and used for the outside space or more general application scenarios. However, a convenient model for indoor location-based services is missing. Therefore, this chapter explains two standardization efforts, which might have impact on the future development of standardization for indoor location-based services. However, there is still not enough interest in providing location-based service data, and the process of generating this data is still prohibitively complex. This leads to a fragmented set of data, because any company providing a service is modeling the minimal amount of information for their actual service and this data is not provided to other services, as it is of high value for the company.

This leaves the world with two extremes: on the one hand, there are standards, which try to define data formats, which are useful for many applications. On the other hand, there are plenty of applications, which do not yet adopt these standards, as they are complicated and the needed data is seldom available.

Consequently, the big open question for indoor location-based services and environmental models will be about who wins the discussion: is it possible to express indoor environmental information flexible enough without increasing modeling cost beyond a reasonable limit? And will there be enough entities sharing this data or will the data be too valuable?

4.6.1 GML and CityGML

Generalizing from indoor location-based services, one can consider environmental models as problems of geometrical representation. For flexible representation of geometric objects, the geometric markup language (GML) provides an extensible framework. As GML is defined as an XML grammar, it provides flexible extension and decouples data representation and actual data from each other. While GML is one of the most important standardizations for geodata, the representation of geometry is limited to constructive vector graphics. The most important objects to represent geometry are given by points, lines, and polygons. Three-dimensional objects are usually patched together from polygons just as is common in 3D modeling systems. With respect to indoor location-based services, the most important weakness of GML is given by the lack of support for raster graphics. As a consequence, available map information has to be transformed into a vector graphics representation, and some data files will become infeasibly large and complex.

A further step of standardization is given by the CityGML application schema for GML. It extends the GML standard by fixing a way of describing multi-floor buildings including some of their functionalities. However, this model has been defined in order to be able to model cities and not in order to be able to provide indoor location-based services to mobile devices. Therefore, topological information is not modeled in CityGML though it can be implicitly contained. Furthermore, the main philosophy behind the development of CityGML was to find a model that can be useful to more than one application domain. It is not clear, whether extensions of CityGML for indoor location-based services would provide useful insights for other application domains.

4.6.2 Indoor OSM

Complementary to the large standardization bodies, a lot of entities are working on practicable and simple map representations and environmental models for buildings. A promising approach of this category is given by IndoorOSM. The OpenStreetMap (OSM) is a large, crowdsourced, free map of the world. It is based on a simple vector graphics data model containing nodes, ways, and relations. These three objects are used to model the complete world. A node represents a location and can be thought of as a spatial point. A way is composed of several nodes and leads to a polygonal linestrip. This is used to model all one-dimensional objects. A relation can then be defined between ways to group several line strips. This is used either to generate a two-dimensional object (e.g., a polygon) or to group several lines into one logical object. In order to facilitate extensions, a free tagging approach is integrated. Tags can be added to the OSM data model as a key-value pair. Some of the keys have been specified or converged to a common meaning; some semantics are still represented by multiple different tags.

With this flexible model, the original OSM project is able to model the world in quite large detail as parts of the community add detailed information for specific user groups. From the central dataset, a renderer extracts the relevant information, which is actually relevant to the user and generates map images from this.

The IndoorOSM project provides an extension of the OSM data model in order to enable building modeling. Therefore, solely existing OSM elements are being used. This approach is quite promising, as the complexities of specific buildings can be modeled by the free tagging approach. Furthermore, the level of detail of mapping is not actually defined. The building objects, for which some standardization has been proposed, are given by specifying buildings, floors, doors, and vertical passages. A building is defined to be a relation containing other relations or elements. Commonly, a building consists of several relations each representing a floor. Possibly, some point objects (e.g., nodes) are also added to the building such as the entrance. These floor relations are composed of building parts, which can represent rooms or walls or other elements. A short list of tags has been defined in order to model the most important properties in the same way. One of the most important definitions is given by the way in which vertical connections between floors are modeled. In order to allow for navigation applications, vertical passages connect door objects, and in this way, a door object is used to represent most of the topological and navigational structure of a building.

4.7 Summary

Environmental models provide environmental information to indoor location-based services without which these services could not provide sensible information to the user. Due to the increased complexity of both indoor spaces and indoor position determination, the use of environmental models is far more important as compared to outdoor location-based services as they are also an integral part of reliable position determination. With respect to coordinate systems, coordinate transformation and map projections, which are essential for scaling location-based services around the globe, further information is given in books on global positioning including Groves and Hofmann-Wellenhoff [3,4]. The orthogonal representations of geometry by vector maps and raster maps provide additional complexity and imply the need for complex decisions, data structures, and algorithms being able to deal with possible inconsistencies. For these topics, classical textbooks on computational geometry can serve as additional sources of information [2]. It is very important that building modeling enters a new decade in which classical approaches like GML and CityGML, which are difficult to handle for mobile devices due to their size and complexity, are merged with minimalistic approaches such as occupancy grids in which calculations with mobile devices are possible without investing a lot of computation and battery power.

In summary, a concise, flexible, and widely adopted description language for indoor environmental models is missing though the topic is quite important and has been discussed for long [1,5,8].

References

1. Becker, C., Dürr, F.: On location models for ubiquitous computing. Pers. Ubiquitous Comput. 9(1), 20–31 (2005)
2. De Berg, M., Van Kreveld, M., Overmars, M., Schwarzkopf, O.C.: Computational Geometry. Springer, Berlin (2000)
3. Groves, P.D.: Principles of GNSS, Inertial, and Multisensor Integrated Navigation Systems. Artech House, London (2013)
4. Hofmann-Wellenhof, B., Legat, K., Wieser, M.: Navigation – Principles of Positioning and Guidance. Springer, Wien (2003)
5. Kolbe, T.H., Gröger, G., Plümer, L.: Citygml: interoperable access to 3d city models. In: Geo-Information for Disaster Management, pp. 883–899. Springer, Berlin (2005)
6. Proj.4 - cartographic projections library. Online (2014). http://trac.osgeo.org/proj/
7. Rigaux, P., Scholl, M., Voisard, A.: Spatial Databases: With Application to GIS. Morgan Kaufmann, San Francisco (2001)
8. Stoffel, E.P., Lorenz, B., Ohlbach, H.J.: Towards a semantic spatial model for pedestrian indoor navigation. In: Advances in Conceptual Modeling—Foundations and Applications, pp. 328–337. Springer, Berlin (2007)

Chapter 5
Position Refinement

> *But since all our measurements and observations are nothing more than approximation to the truth, the same must be true of all calculations resting upon them, and the highest aim of all computations made concerning concrete phenomena must be to approximate, as nearly as practicable, to the truth.*
>
> Karl Friedrich Gauss

The inference of the position of a mobile device is usually discussed first for a single instance in time: at a given time t, some measurements are made, and the most probable location of the mobile device is to be detected. This ignores two useful facts: in most applications, the position at a specific instance in time is similar to the position a short time later. Two consecutive positions determined by a positioning system are therefore not independent from each other. Secondly, however, the errors are when two measurements at two different instances in time have been made for the same or a similar location, then the errors involved in the measurements will be partly uncorrelated. Therefore, a better estimate can be given by considering both measurements together than by treating them isolated.

In order to exploit both facts, some algorithms have been proposed, which model the relation between the current state of the system with previous states. Most often, the previous states are kept in an aggregated form such that the algorithms work with a constant amount of memory and information.

5.1 Least Squares Estimation with Correlation

In positioning and navigation systems, different errors are often correlated. Consider a simple inertial navigation system: acceleration measurements with zero-mean Gaussian noise are used to estimate velocity by calculating the integral over time. The estimation error of velocity then grows linearly with time. Hence, the position error, calculated as the integral of the velocities, will have quadratically increasing errors. However, these errors are all correlated. A larger velocity estimation error leads to a faster growth of position error.

© Springer International Publishing Switzerland 2014
133
M. Werner, *Indoor Location-Based Services*, DOI 10.1007/978-3-319-10699-1_5

For this situation, it is useful to extend the least square estimation to incorporate for this correlation of state space variables. For a set of random variables $X_1 \ldots X_n$, this correlation can easily be expressed by the $n \times n$ matrix given by pairwise covariances

$$\Sigma = \Sigma_{i,j} = \mathrm{cov}(X_i, X_j) = \mathbb{E}\left[(X_i - \mu_i)(X_j - \mu_j)\right].$$

This is a generalization of variance to account for different variables. The diagonal entries simply recover the variance

$$\mathrm{var}(X_i) = \mathrm{cov}(X_i, X_i).$$

This covariance matrix can be used to weight different measurements. These weights should express the degree of confidence that can be placed into the individual variables. It has shown to be useful to use the inverse of the covariance matrix for this weighting. This is easily motivated by the following simple example: consider a set of pairwise uncorrelated variables. The covariance matrix becomes a diagonal matrix consisting of the variance of each variable. The inverse matrix is then given by inverting the diagonal elements; hence, a large variance leads to a small value in the inverted covariance matrix, a small variance to a large value. Hence, measurements with low expected error have larger weight compared to highly erroneous measurements.

Problem 5.1 Given a set of measurements b with Gaussian noise of covariance matrix R, find the value x in the linear matrix equation $Ax = b$, which minimizes the weighted residual square sum

$$b^T R^{-1} b = (y - Ax)^T R^{-1}(y - Ax) \rightarrow \min.$$

A sequence of measurements for a navigation system using, for example, lateration (Sect. 3.1.2) leads to a sequence of overdetermined systems of linear equations of the form

$$Ax = b.$$

This can also be seen as a single overdetermined system of linear equations in random variables. Let R denote the covariance matrix of the observation sequences. Then we use the inverse of R as a weighting for the influence of each single equation on the solution

$$b^T R^{-1} b = (y - Ax)^T R^{-1}(y - Ax) \rightarrow \min.$$

Performing this minimization leads to the equivalent set of normal equations

$$A^T R^{-1} Ax = A^T R^{-1} b$$

and hence to an estimator

$$x = \left(A^T R^{-1} A\right)^{-1} A^T R^{-1} b.$$

Note the similarities to Eq. (3.1) in Chap. 3, where unweighted least squares was introduced in detail. Unweighted squares is easily recovered by taking R and hence R^{-1} to be the identity matrix.

It is known from statistics that this estimator is the best linear unbiased estimator for the given cases. It generalizes over the classical least square approach in that it does not need statistically independence and inherently incorporates the dynamics of the observations.

The law of error propagation can be used to find the covariance matrix of the estimate x:

$$P = \left(A^T R^{-1} A\right)^{-1}.$$

This completes a simple algorithm for estimating x in cases, where the measurements b are correlated.

Algorithm 1 Weighted least square algorithm

1: Calculate R^{-1}
2: $P := \left(A^T R^{-1} A\right)^{-1}$
3: $x := P A^T R^{-1} b$
4: **return** x

5.2 Recursive Least Squares Estimation

The previous section showed a method of dealing with a quite general class of optimization problems: linear problems with correlated variables. However, the fact that some variables are uncorrelated can be used to reduce the complexity of the system. Assume a partition of the measurements b_i into sets such that two observations from different sets are uncorrelated. Then the covariance matrix R obtains a diagonal block structure

$$R = \begin{pmatrix} R_0 & 0 & \cdots & 0 \\ 0 & R_1 & \cdots & 0 \\ \vdots & \vdots & \vdots & \vdots \\ 0 & \cdots & & R_n \end{pmatrix}$$

The central result behind recursive least square estimation is the fact that the solution x_j for the jth block of equations can be calculated from the results of the previous block with indices $j - 1$ alone. This can be either rigorously shown by calculations. However, there is a clear motivation behind: as each intermediate estimate x_k along with its covariance P_k is a summary of all previous measurements, the following x_{k+1} can be calculated from less data than all measurements $b_0, \ldots, b_{k+1}$. Fortunately, this is the case. Algorithm 2 gives a recursive description of estimating x_n accessing only data and results from the current block j and the previous block $j - 1$. This gives a sequence of estimate x_j for each block of the matrix.

Assuming that the noise of a discrete sampling of a time-dependent signal, one can use this procedure as an online algorithm for estimating a dynamic linear system from incoming measurements. This would be a major change of attitude compared to classical least square estimation, as the observed state x_k need not be constant anymore. The matrices K_j are called gain matrices, as they can independently be

---

Algorithm 2 Recursive weighted least square algorithm

1: Calculate R
2: **for** $j = 1 \to n$ **do**
3:     $K_j := P_{j-1} A_j^T (A_j P_{j-1} A_j^T + R_j)^{-1}$
4:     $x_j := x_{j-1} + K_j \left(b_j - A_j x_{j-1} \right)$
5:     $P_j := (1 - K_j A_j) P_{j-1}$
6: **end for**
7: **return** x_n

---

derived from varying the gain matrices of a general linear filter term for minimizing the mean squared error. This minimization leads to the Wiener–Hopf equations; see [6] for details.

5.3 Discrete Kalman Filtering

The recursive approach to weighted least square estimation leads to a straight-forward treatment of time-varying discrete signals. Assuming that the individual noise terms for measurements taken at different instances in time t_k are pairwise independent, one can employ the recursive estimation approach to estimate the state x_k from all previous measurements $b_0, \ldots, b_k$. As explained in the previous section, this is even possible without holding all measurements $b_0, \ldots, b_k$ and intermediate states x_k in memory. The recursive least squares approach allows for a constant memory and constant runtime filtering providing the optimal result comparable to holding everything in memory and solving the complete set of linear equations using weighted least squares.

Applying recursive least squares in order to use measurements taken at several different timestamps is a first step towards Kalman filtering. Discrete Kalman filtering is best derived as an extension to the recursive least squares method. The recursive least square method consists of a loop which needs access to some data from the previous iteration and A_j and R_j from the current iteration.

While the recursive least squares approach calculates in every iteration j with state from the past time instant $j - 1$, the question is raised, whether this state can be updated reflecting the elapsed time between both iterations. This is done in a Kalman filter. The Kalman filter is obtained from the recursive least squares algorithm by replacing the previous iteration state x_{j-1} and the associated covariance P_{j-1} by a prediction of the state and covariance using a linear system model of the form

$$x_{j+1} = B_j x_j + v_j$$
$$v_j \sim \mathbf{N}(0, Q)$$

where B is an $n \times n$ matrix called transition matrix and v represents a zero-mean Gaussian variable with covariance matrix Q called system noise.

It is now customary to represent the complete set of equations defining the Kalman filter as three blocks:

- *Gain computation* in which the contribution of measurements and prediction models are weighted based on the current uncertainties
- *Measurement update (correction)* in which a newly arriving measurement is integrated into the current state of the system
- Time update (prediction) in which the system state is predicted into the future based on the previous system state

Introducing new sizes $\tilde{P}_{j+1}$ for the predicted covariance at future timestamp t_{j+1} and $\tilde{x}_{j+1}$ for the predicted state, one can calculate as follows:

$$\tilde{x}_{j+1} = B_j x_j$$
$$\tilde{P}_{j+1} = B_j P_j B_j^T + Q_j.$$

Substituting these sizes as appropriate into the algorithm of recursive least squares estimation, that is, replacing P_{j-1} by $\tilde{P}_j$ and x_{j-1} by $\tilde{x}_j$, leads to the following formulation of the Kalman filter:

The matrix K_j is called Kalman weight, and with two extreme examples, it can be easily seen that it actually represents some weighting between the measurement model and the prediction model. Assume that a system is able to directly observe the state. Then the matrix A is the identity matrix. In this situation, the formula for the Kalman gain computation degenerates to

$$K_j = \tilde{P}_j (\tilde{P}_j + R_j)^{-1}.$$

Algorithm 3 Kalman filter algorithm

1: **for** $j = 1 \to n$ **do**
2:     *Compute Kalman Gain*
3:     $K_j := \tilde{P}_j A_j^T (A_j \tilde{P}_j A_j^T + R_j)^{-1}$
4:     *Measurement Update (Correction)*
5:     $x_j := \tilde{x}_j + K_j (b_j - A_j \tilde{x}_j)$
6:     $P_j := (1 - K_j A_j) \tilde{P}_j$
7:     *Time Update (Prediction)*
8:     $\tilde{x}_{j+1} := B_j x_j$
9:     $\tilde{P}_{j+1} := B_j P_j B_j^T + Q_j$
10: **end for**
11: **return** x_n

Consider now the case that the measurements are very accurate. This would be reflected by an almost zero measurement covariance matrix $R_j \approx 0$. In this case $(A = K_j = 1)$, the Kalman gain matrix K_j approaches the identity matrix. In this case, the measurement update reads

$$x_j \approx \tilde{x}_j + \left(b_j - \tilde{x}_j\right) = b_j.$$

In the opposite case, assume that the prediction accuracy is very high reflected by an almost zero time update covariance matrix $\tilde{P}_j \approx 0$. In this case, the Kalman gain computation results in

$$K_j \approx 0,$$

and the measurement update leads to

$$x_j \approx \tilde{x}_j.$$

This simplified example showed that for very precise measurements, the Kalman equations use only measurement information and that for highly precise prediction estimates, the filter prefers predicted information. This is the reason, why the matrix K is often called gain. Figure 5.1 depicts the information flow for classical Kalman filter information. Note that reasonable initial values have to be assigned to the different covariance matrices and state variables in order to apply a Kalman filter in reality.

The Kalman filter algorithm is an optimal algorithm for estimating the state in the situation where the errors are Gaussian and the models are actually linear. This means that no other algorithm can be any better in this linear and Gaussian environment. However, many environments are not linear, and this motivates an extension to the Kalman filter for nonlinear equations.

Fig. 5.1 Information flow
inside the Kalman filter
algorithm

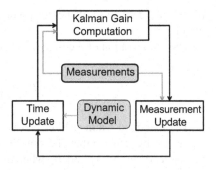

5.4   The Extended Kalman Filter

The extended Kalman filter is just like the Kalman filter, an algorithm for estimating the state of a system out of a sequence of measurements. However, the equations do not have to be linear anymore; they must only be differentiable. Assume that the behavior of the system is modeled using the following two differentiable, possibly nonlinear equations.

Let the relation between state and observations be given by the system equation

$$b_j = \alpha_j(x_{j-1}) + v_{j-1}$$
$$v_{j-1} \sim \mathbf{N}(0, R_{j-1})$$

where v_{j-1} represents zero-mean Gaussian random noise with covariance matrix R_{j-1}. Further, let the behavior of the system with respect to time be given by

$$x_{j+1} = \beta_j(x_j) + w_j$$
$$w_j \sim \mathbf{N}(0, Q_j)$$

where w_j represents zero-mean Gaussian random noise with covariance matrix Q_j.

Though the system is quite similar to the Kalman filter, the Kalman filter algorithm cannot be applied due to the nonlinearity of α and β. For the extended Kalman filter, one uses now a linearization of both equations. The linearizations are given by calculating the Jacobian matrix of α and β. Let therefore e_k denote the k-th basis vector of $\mathbb{R}^n$. The Jacobian of $\alpha_j : \mathbb{R}^n \to \mathbb{R}^n$ evaluated at the vector $\tilde{x}_j$ can be written as

$$A_j = \left(\frac{\partial \alpha_{j,i}}{\partial e_k}\right)_{i,k} (\tilde{x}_j).$$

In this equation the partial derivations

$$\frac{\partial \alpha_{j,i}}{\partial e_k}$$

denote the derivation of the ith function $\mathbb{R} \to \mathbb{R}^n$ of α_j with respect to the kth parameter.

The linearization of the function β_j can be written in a similar way:

$$B_j = \left(\frac{\partial \beta_{j,i}}{\partial e_k} \right)_{i,k}.$$

The extended Kalman filter can now easily be derived from the Kalman filter algorithm by using either the non-linear functions α_j and β_j where possible or their linearizations A_j and B_j where needed. For the gain computation, the matrix K_j is calculated using the linearization A_j. For the measurement update, x_j is calculated using α_j, and the associated covariance matrix P_j is calculated from the linearization A_j. For the prediction step, $\tilde{x}_{j+1}$ is calculated from β_j, while the associated covariance matrix $\tilde{P}_j$ is calculated from the linearization B_j.

Putting this all together leads to the final Algorithm 4.

Algorithm 4 Extended Kalman filter algorithm

1: **for** $j = 1 \to n$ **do**
2:      *Compute Kalman Gain*
3:      $K_j := \tilde{P}_j A_j^T (A_j \tilde{P}_j A_j^T + R_j)^{-1}$
4:      *Measurement Update (Correction)*
5:      $x_j := \tilde{x}_j + K_j \left(b_j - \alpha_j(\tilde{x}_j) \right)$
6:      $P_j := (1 - K_j A_j) \tilde{P}_j$
7:      *Time Update (Prediction)*
8:      $\tilde{x}_{j+1} := \beta_j(x_j)$
9:      $\tilde{P}_{j+1} := B_j P_j B_j^T + Q_j$
10: **end for**
11: **return** x_n

Note that this algorithm can often be combined with recursive least square estimation inside each time epoch. For each time epoch, there is a least square estimation to be done. If the measurement variables are partly independent from each other in the sense that the covariance matrix R has a block diagonal structure, then again computations can be performed recursively. This has two good effects: First of all, the size of the individual matrices is smaller, and as most matrix operations have nonlinear running time, this saves a lot of computation time. Hence, it makes the algorithm faster. And secondly, the matrix inversion of these smaller matrices with fewer zero entries gains numerical stability. Moreover, the individual measurements need not be processed at once: if some measurements are missing, the other blocks can still be calculated. This can become very useful in cases of varying availability or update rates of the individual measurements. In this sense, the extended Kalman filter, and the Kalman filter as well, can be driven by incoming data.

5.5   Particle Filtering

The Kalman filter introduced the central idea of using two models to estimate
the state of a system: a prediction model, which describes the evolution of the
system state over time, and a measurement model, which relates possibly noisy
measurements to the current state estimate. For the Kalman filter, these models
have both been specified in deterministic form. The time update equation calculates
the future state of the system based on the current state, and the measurement
update equation calculates the new state out of the current state and an incoming
measurement. For particle filtering, these are given in probabilistic form. The goal
of a particle filter is to estimate the posterior probability distribution of the state at
a specific point in time using all measurements $b_0 \ldots b_j$ of the past. From such a
probability distribution, the most probable state can actually be extracted, and the
system also provides an estimate of the accuracy of this state estimate.

For particle filter, the recursive nature derived for the Kalman filter shall be kept
intact. In general, the time update equation will translate and spread the state. The
longer no measurements are included into the state estimate, the less accurate does
the state represent the true state. The measurement update does the reverse. When
measurements are integrated, the estimate of the state becomes less uncertain, and
the probability density function (pdf) of the state will be more compact.

Though particle filters can be applied to a lot of different problems, we will
concentrate on the case of the nonlinear Bayesian tracking problem. In analogy to
the description of the Kalman filter, let the state sequence be given as

$$b_j = \alpha_j (x_{j-1}, v_{j-1}) \tag{5.1}$$

where α_j is a possibly nonlinear function of the state and v_{j-1} is an i.i.d.
measurement noise sequence.

Further, let the behavior of the system with respect to time be given by

$$x_{j+1} = \beta_j (x_j, w_j) \tag{5.2}$$

where β_j is a possibly nonlinear function of the state and w_j is an measurement
noise sequence. Let us assume that w_j are independent and identically distributed
(i.i.d.).

The task is now to find estimates x_k given all available measurements $b_{1..k} = \{b_1 \ldots b_k\}$. In a probabilistic sense, it is required to construct the pdf

$$p(x_k | b_{1\ldots k})$$

of the state x_k given all measurements. Assuming that an initial probability density
$p(x_0)$ is given, the probability density function can be given by the two steps
prediction and correction in a recursive way:

Prediction: Assume that the pdf $p(x_{k-1}|b_{1...k-1})$ is available to the system. Then the prediction model in Eq. (5.2) can be used together with the Chapman–Kolmogorov equation to determine

$$p(x_k|b_{1...k-1}) = \int p(x_k|x_{k-1})p(x_{k-1}|b_{1...k-1})dx_{k-1}.$$

The factor $p(x_k|x_{k-1})$ is described by Eq. (5.2).

Correction: Assume that a new measurement b_k has become available to the filter. Then the rule of Bayes can be expressed as

$$p(x_k|b_{1...k}) = \frac{p(b_k|x_k)p(x_k|b_{1...k-1})}{p(b_k|b_{1...k-1})}.$$

The measurement model of Eq. (5.1) trips into this equation twice: once in the factor $p(b_k|x_k)$ in the nominator and again in the denominator, which can be expressed as

$$p(b_k|b_{1...k-1}) = \int p(b_k|x_k)p(x_k|b_{1...k-1})dx_k.$$

Note that the Kalman filter perfectly fits into this framework and that all integral equations can be spelled out in terms of covariances and matrix multiplications leading to the very same recursive algorithm as given in Algorithm 3.

While this formulation is quite elegant in theory, it is mostly useless in practice without comments on how to compute the involved integrals. The following two subsections present methods with which these calculations can be performed in special cases.

5.5.1 Grid-Based Methods

When the state space can be split into N cells $\{x_k^i, i = 1...N\}$, then a grid-based method can be used to approximate the posterior density. In this case, the probability densities can be written as finite, weighted sums of Dirac distributions. Using these simplifications, the posterior pdf at time $k-1$ can be rewritten as

$$p(x_{k-1}|b_{1...k-1}) \approx \sum_{i=1}^{N} w_{k-1|k-1}^i \delta(x_{k-1} - x_{k-1}^i),$$

and the prediction equation becomes

$$p(x_k|b_{1...k-1}) \approx \sum_{i=1}^{N} w_{k|k-1}^i \delta(x_{k-1} - x_k^i),$$

and the correction equation can be written as

$$p(x_k|b_{1...k}) \approx \sum_{i=1}^{N} w_{k|k}^i \delta(x_{k-1} - x_k^i).$$

The factors w^i can now be calculated for each cell using numerical integration over the area represented by the cell i, and this gives a working algorithm.

5.5.2 Sampling Importance Resampling

The grid-based methods always provide a homogenous approximation of the state space. Therefore, a lot of calculations have to be done for cells, which are currently not relevant. Especially for filtering of location information, the involved probability density functions will usually have small support, and hence, a lot of calculations are performed for cells, which actually do not contribute to the estimation. Therefore, it should be possible to adapt the algorithm to work more in areas, where the pdf is non-zero as compared to the area, where it vanishes clearly.

Sampling importance resampling (SIR) particle filter now flexibilizes the grid in that it approximates the probability density functions using a finite set of N particles. Each particle represents a region of the state space around the particle center and is associated with a weight. These particles together can be seen as a generalized histogram of bins. Each particle represents a bin around its current location, and the weight represents the relative number of elements added to the bin.

The approach to particle filtering is now easy and quite similar to the Kalman filter: each particle is updated in time by an update model which can update each particle or even generate new particles to represent the uncertainty of time update. If, for example, the state contains a direction which is only known up to some standard deviation, the time update can create a set of particles each representing a hypothetical truth given the distribution of possible movement directions. For an incoming measurement, again, each particle is updated with respect to its current state and probability of representing the truth. In this step, it is possible to integrate map matching to reduce the weight of particles which go through walls or have otherwise unlikely behavior inside the map.

As particles can be split in order to correctly model uncertainty in the form of new hypotheses, the system needs to maintain an increasing number of particles over time. Due to computational limitation, the number of particles has to be reduced from time to time by removing particles with a small weight from the current set of particles. When the number of particles is large enough, particle filter produces an optimal estimate. However, this approach can be impractical due to memory and computational limitations. On the contrary, when too few particles are used to represent a complicated probability density function, the error can be arbitrarily large leading to completely wrong results. Fortunately, the calculation of particles is easily parallelizable, and hence, particle filters can be scaled across cores or machines to some extent.

The first particle filter was proposed by Gordon in 1993 [3]. This filter consists of the following three steps: prediction, correction, and resampling.

Prediction: Each particle is passed through the system Equation (5.2):

$$\tilde{x}_{j+1}(i) = \beta_j(x_j, w_j(i)).$$

Therefore, an explicit noise sample $w_j(i)$ is drawn from the known distribution $p(v_j)$ of v_j. In this way, the noise term will introduce uncertainty as the result is parametrized randomly for each particle.

Correction: For each incoming measurement, evaluate the likelihood of each particle and normalize across all particles in order to get a normalized weight for all particles

$$w_i = \frac{p(b_j | \tilde{x}_j(i))}{\sum_{k=1}^{N} p(b_j | \tilde{x}_j(k))}.$$

The calculation of the probability is based on using the known statistics of the measurement model (5.1).

Resampling: From the set of N particles, sample a new set of N particles from the discrete distribution built from the weighted particles $(w_i, \tilde{x}_j(i))$.

The great strength of this approach is the fact that no assumptions about Eqs. (5.2) and (5.1) such as linearity or differentiability are needed. The only preconditions to this algorithm are that:

- An initial distribution $p(x_1)$ is available for sampling.
- The likelihood $p(b_j | x_j)$ is known and can be evaluated as a function.
- $p(w_k)$ is available for sampling.

Another strength is that the algorithm is very easy to implement and that there are fewer numerical pitfalls such as matrix inversion or partial derivatives as compared to the extended Kalman filter. The algorithmic ingredient to easily implementing the resampling step is based on a roulette wheel technique: draw a random sample $u \in (0, 1]$ from a uniform distribution. Then the space $(0, 1]$ is split into consecutive cells of length w_i, and the ith particle is chosen, when u falls into the ith cell given by

$$\sum_{k=0}^{i-1} w_k < u \leq \sum_{k=0}^{i} w_k.$$

Geometrically, this amounts to a roulette wheel, where the size of the i-th slot is given by w_i as depicted in Fig. 5.2.

Fig. 5.2 Roulette wheel approach for resampling from a weighted set

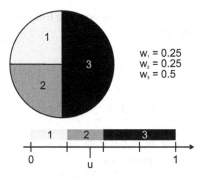

$w_1 = 0.25$
$w_2 = 0.25$
$w_3 = 0.5$

Algorithm 5 gives a particle filter in pseudocode as a hint towards implementing it yourself. The particle filter runs for timesteps $j = 1 \ldots T$ and contains N particles, which have been drawn from some distribution for initialization. The functions α and β denote the measurement update and time update functions just as before. The current state x has been initialized. We assume that at any time instant j, a new measurement b_j is available.

Algorithm 5 Particle filter

1: **for** $j = 1 \to T$ **do**
2:     $b_j :=$ getMeasurement()
3:     **for** $i = 1 \to N$ **do**
4:         *Update Particle*
5:         $\tilde{x}_j(i) := \beta(x_j(i), t)$
6:         $\tilde{b}_j(i) = \alpha(\tilde{x}_j(i))$
7:         *Weight Measurement vs. Expected Measurement*
8:         $w_i = p(b_j | \tilde{x}_j(i), \tilde{b}_j(i))$
9:     **end for**
10:    *Normalize*
11:    **for** $i = 1 \to N$ **do**
12:        $w_i = \frac{w_i}{\sum_k w_k}$
13:    **end for**
14:    *Resample*
15:    **for** $i = 1 \to N$ **do**
16:        $u = \mathbf{rand}(0, 1)$
17:        Find p with $\sum_{k=0}^{p-1} w_k < u \leq \sum_{k=0}^{p} w_k$
18:        $x_i = \tilde{x}_p$
19:    **end for**
20:    $x_e = \frac{\sum_k x_k}{N}$
21: **end for**

In order to have a completely concrete representation of at least one particle filter, we will now fill in the line 8 in Algorithm 5 for a concrete example. This example was also given in the original work of Gordon [3].

Example 5.1 Consider the following one-dimensional example for a particle filtering problem:

$$\alpha(x) = \frac{1}{20}x^2 + \mathcal{N}(0,1) \tag{5.3}$$

$$\beta(x,t) = \frac{1}{2}x + 25\frac{x}{1+x^2} + 8\cos(1.2(t-1)) + \mathcal{N}(0,10). \tag{5.4}$$

For an initial state of $x = 0.1$, Fig. 5.3 depicts a possible realizations of these random processes.

Due to the Gaussian system noise and the Gaussian measurement noise, line 8 of the particle Filter Algorithm 5 can be spelled out for the Gaussian case. The weights are defined to be the probability of the given observation b_j given the actual predicted state of that particle. In other words, when we observe a location b_j with Gaussian error of variance R, the probability of getting $\tilde{b}_j(t)$ is given by

$$w_i = \frac{1}{\sqrt{2\pi R}} \exp\left(\frac{-(b_j - \tilde{b}_j(t))^2}{2R}\right).$$

This completes a particle filter algorithm in a concrete case. In different applications, the functions α and β will vary, and most importantly, the calculation of the weights w_i based on incoming measurements will be different depending on the statistics of the observation error. Having a Gaussian observation error is, however, one of the most important cases.

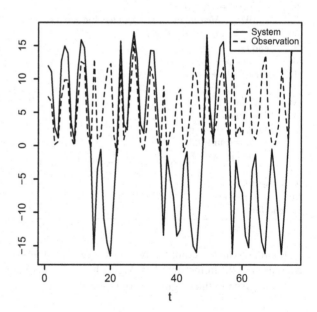

Fig. 5.3 The dynamic process given by α and β

Now that we have seen a one-dimensional example of filling in all details of the particle filter to complete an implementation, we should consider another example more tailored to the domain of positioning and location-based services.

Example 5.2 Therefore, let us consider a two-dimensional filtering scenario in which particles represent a pose consisting of coordinates and orientation. Therefore, we assume that we can directly observe the moving angle of a mobile entity up to zero-mean Gaussian noise. This type of information could be provided from a digital compass in reality. Let us further assume that the mobile entity is moving with constant speed. Let now a particle store the x coordinate, the y coordinate, and the heading of the mobile device

$$\alpha : \begin{pmatrix} x \\ y \\ \phi \end{pmatrix} \rightarrow \begin{pmatrix} x \\ y \\ \phi + \mathcal{N}(0, 0.3) \end{pmatrix}.$$

This expresses the probabilistic relation between incoming measurements and the truth given by normally distributed influence on the angle. The position is kept untouched in the measurement model in this case. The time update model is given by expressing that the mobile device shall be moving into the direction given by its current heading; in other words

$$\beta : \begin{pmatrix} x \\ y \\ \phi \end{pmatrix}, \Delta t \rightarrow \begin{pmatrix} x \\ y \\ \phi \end{pmatrix} + \Delta t \begin{pmatrix} \cos(\phi) \\ \sin(\phi) \\ 0 \end{pmatrix}.$$

The weighting of particles with respect to an incoming measurement ϕ_j of the angle is again calculated from a normal distribution according to the formula

$$w_i = \frac{1}{\sqrt{2\pi R}} \exp\left(\frac{-(\phi_j - \tilde{\phi}_j(t))^2}{2R} \right),$$

for each particle. Figure 5.4 depicts the particle filter working: the left is the starting time, where a normal distribution around the point $(0, 0, 0)$ has been sampled to initialize particles. Then, angles increasing from 0 to 2π are incoming as new measurements for the individual times. After half of the time, the filter is depicted in the middle, and the final state is depicted in the right of Fig. 5.4. Here you can see one very typical effect of filtering: the movement model assumes linear movement along the tangent to the circle in this case. Hence, the estimates drift away from the circle, and the circle does not close. This error is to be expected as the chosen motion model and particle representation are not sufficient to model this type of nonlinear movement.

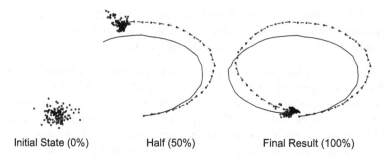

Initial State (0%)   Half (50%)   Final Result (100%)

Fig. 5.4 A particle filter filtering a circular rotation using only heading information and a linear movement model

5.6 Summary

Refining the accuracy of positioning systems for indoor location-based services is essential due to the comparably large error of positioning systems and the comparably high accuracy needs. This chapter has given an introduction to the two most important approaches to refining positioning and integrating additional sources of information into the positioning system. Both approaches have in common that they try to extract information from the time-domain evolution of signals: either by assuming that some error components are independent of time and therefore cancel each other over time or by integrating additional information in a suitable representation of the current belief of a navigation system. For the first approach, the recursive filtering in general and the Kalman filter with its extensions have been widely applied due to their high performance and theoretical basement. For the second approach, particle filter have been used often, as it is easy to change the probabilistic belief of this data structure by updating weights associated with particles by external information. In this way, for example, floorplans can be used to erase or push back particles when they cross walls or occasionally available sensors can be integrated without tracking their state for the complete time as in the Kalman filter case.

5.7 Further Reading

The area of filtering has been widely recognized and is a very active area of research. A nice overview on particle filter and Kalman filtering and their individual relations is presented in a survey by Chen [2]. More introductory in nature and focused on applications to navigation and tracking are sections from the book of Hofman-Wellenhof, Legat, and Wieser *Navigation* [5, Chap. 3.6] and from Paul D. Groves

Principles of GNSS, Inertial, and Multisensor Integrated Navigation Systems [4, Chap. 3, 16]. A very concise discussion of the various filters is given by a tutorial of Arulampalam et al. [1].

References

1. Arulampalam, M., Maskell, S., Gordon, N., Clapp, T.: A tutorial on particle filters for online nonlinear/non-gaussian bayesian tracking. IEEE Trans. Signal Process. **50**(2), 174–188 (2002). doi:10.1109/78.978374
2. Chen, Z.: Bayesian filtering: from Kalman filters to particle filters, and beyond. Statistics **182**(1), 1–69 (2003)
3. Gordon, N., Salmond, D., Smith, A.: Novel approach to nonlinear/non-gaussian bayesian state estimation. In: Proceedings of the Institute of Electrical Engineering, vol. 140, pp. 107–113 (1993)
4. Groves, P.D.: Principles of GNSS, inertial, and multisensor integrated navigation systems. Artech House, London (2013)
5. Hofmann-Wellenhof, B., Legat, K., Wieser, M.: Navigation – Principles of Positioning and Guidance. Springer, Wien (2003)
6. Sorenson, H.W.: Least-squares estimation: from gauss to Kalman. IEEE Spectr. **7**(7), 63–68 (1970)

Chapter 6
Trajectory Computing

> *The most basic question is not what is best, but who shall decide what is best.*
>
> Thomas Sowell

Location-based services are most often based on a single location due to various reasons. One of the most important reasons is given by the simple nature of location. A fundamental result from math states that well-behaved functions for computing the distance between points in normed vector spaces are essentially the same. In every finite-dimensional vector space, two different norms are equivalent in the sense that they provide the same geometric principles to the space.

However, location-based services are usually producing time series of location. This is especially true for proactive services, tracking applications, and navigation applications. However, computation with this data is quite hard as there is no result comparable to the norm equivalence for finite-dimensional vector spaces. Therefore, a lot of different distance measures for trajectories have been proposed, which all have their advantages and disadvantages for a given application scenario.

An illustrative example for two completely different types of trajectory distances is given by comparing handwriting recognition with trajectory comparison of GPS traces. While in the handwriting case, the task is given by finding similar shapes defining different letters without considering the spatial location of the letter on a sheet of paper, the GPS case is not that much interested in different shapes than in the absolute deviations between two trajectories.

One reason to extend location-based services to using spatial trajectories is given by new possibilities to overcome inaccuracies and by new possible services and more reliable predictions of future location, action, and relevant information. In order to overcome inaccuracies, trajectories provide strong power under the assumption that at two different timestamps, the location errors are statistically independent. Note that we have used this fact already for recursive least squares estimation (cf. Sect. 5.2). In this sense, we already discussed some trajectory-based services with respect to filtering. In this chapter, we want to exploit the additional information contained in trajectories directly and will therefore discuss *trajectory-based services*.

© Springer International Publishing Switzerland 2014
M. Werner, *Indoor Location-Based Services*, DOI 10.1007/978-3-319-10699-1_6

6.1   The Process of Trajectory Computing

Trajectory computing is best envisioned as a process consisting of different steps. For each of these steps, different algorithms can be used for different results. Still, most applications of the domain follow this general pattern.

The first step in typical trajectory computing applications is given by *data collection*. In the data collection stage, each user or device provides sequences of sensor data such as location to a database, which stores this data as is. We can assume that all clients providing sensor data share a sufficiently synchronized common time, for example, from GPS or the cellular network.

Once this data collection is completed, typically *preprocessing* steps are introduced. Three common preprocessing tasks are resampling, outlier rejection, and data reduction. As the sampling times of different devices are usually different, resampling is often applied by interpolating location at specific equal sampling times for all trajectories in the database. Moreover, some trajectory computing algorithms are quite sensitive to outliers, and hence, simple outlier rejection schemes can already be integrated in this step. This is especially sensible, because after resampling, outliers will affect more than one sample of a trajectory and might lose their property of being very local. Furthermore, data reduction can be applied such as the Douglas–Peucker algorithm suppressing points for which the trajectory does not change too much when left out. The third part of the trajectory computing process is given by *trajectory analysis*. This part can itself be multistaged and produce intermediate results. It aims for providing insight into the contents of the database and results derived from it.

6.2   Trajectories

A trajectory is commonly defined to be given as a finite sequence of samples of location, that is, as a sequence of pairs (t, l), where $t \in \mathbb{R}$ denotes the time and $l \in \mathbb{R}^n$ denotes a location represented by a finite-dimensional vector. For simplicity, we will assume that the Euclidean norm can be used to compare these location vectors. This is the natural representation of a trajectory as it is generated. At different points in time, which need not follow any pattern, a location is added to a trajectory.

A more general definition of a trajectory removes the sampling nature of computer systems. Thereby, it becomes possible to express the continuous nature of trajectories. In this setting, a trajectory α is a *continuous* map:

$$\alpha : \mathbb{R} \to \mathbb{R}^n.$$

In that regard, the continuity of α means that changing the time parameter for only a small amount results in changing the location also for a small amount. In other words, location does not jump in time. Some of the most important distance functions for trajectories are based on this more general definition; however, for computational systems, a more accessible representation, still more general than the sampling representation, is introduced: trajectories in computer systems are most often represented as polygonal curves:

A *polygonal curve* in $\mathbb{R}^n$ is a continuous map:

$$A : [0, 1] \rightarrow \mathbb{R}^n$$

together with a discrete set $\mathscr{S} \subset [0, 1]$ such that for two consecutive points[1] $p, q \in \mathscr{S}$, the restriction of A onto the interval $[p, q]$ is affine, that is,

$$A(p + \xi) = (1 - \xi)A(p) + \xi A(q) \text{ for } \xi \in [0, 1].$$

In less formal language, these are sequences of straight line segments connecting consecutive points in $\mathscr{S}$.

Figure 6.1 depicts three example trajectories for the three different definitions. The leftmost representation is given by the points and their ordering indicated by numbers. Note that actual timestamps can be assigned instead of these numbers. As the true trajectory jumps between the two parts, an approximation as a continuous trajectory is depicted in the second figure. The third figure shows a polygonal curve approximation to the real trajectory. Note that the accuracy of a polygonal approximation of a true trajectory is dependent on the number of points used to represent the trajectory. As the memory demand and the computational complexity of algorithms working on trajectory data depends on the number of samples inside the trajectory, an important problem is given by finding representations of

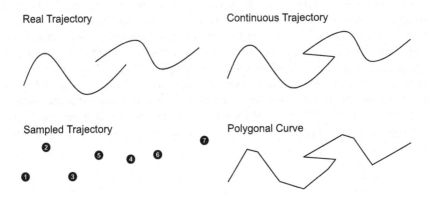

Fig. 6.1 Three types of trajectories

---

[1]Two points p, q are consecutive in $\mathscr{S}$, if the interval (p, q) has empty intersection with $\mathscr{S}$: $(p, q) \cap S = \emptyset$.

trajectories, which approximate the true trajectory with sufficient accuracy while keeping the number of points in the representation low.

Higher-order representations of trajectories are also possible, but usually with additional assumptions about the true trajectory. An example is given by splines, which are a generalization of polygonal curves, which do not only assert equality of location between two affine segments but also equality of some derivatives. In this way, the functions are not only continuous but differentiable and represent a lot of true trajectories better. However, computing with them is even harder than computing with polygonal curves, and due to the inherent errors in real-world trajectories, the assumption of differentiability of observations is sometimes misleadingly restrictive.

In some cases, trajectories are annotated with probabilistic information in trajectory-based services. For example, a positioning system might be able to provide an estimated error distribution together with a position. This leads to the notion of a probabilistic trajectory: a probabilistic trajectory is a trajectory which additionally stores a probability distribution at each location.

The next section will show some of the most important distance functions used to compare different trajectories.

6.3 Trajectory Comparison

As already indicated above, there are several reasonable ways of comparing trajectories. Note that some of the following distance functions are neither a metric nor a norm and, hence, do not provide a reasonable distance in the strong mathematical sense. These distance functions are better seen as scores ranking the similarity of complex objects. The smaller the rank function, the more similar are the arguments. Still, some basic properties are usually kept intact, for example, that the distance of equal objects is zero and that distance is a positive function. The triangle inequality, which states that the length of one edge of a triangle is always shorter than the sum of the length of the others, is not always fulfilled. Most distances in the following are based on a distance between locations. We will usually assume that this distance is given as a norm on the spatial locations from $\mathbb{R}^n$. Note that many approaches actually need less properties and work with similarity scores or metric spaces in a similar way.

6.3.1 Hausdorff Distance

The Hausdorff distance is a classical distance for subsets of metric spaces. It takes an underlying distance function for comparing points and extends it for the comparison of compact subsets of this space. A compact subset of a normed vector space is closed and bounded. Finite polygons are simple examples of closed and bounded subsets of a metric space.

The Hausdorff distance measures the mutual overlap of sets and is constructed from a distance between two points. Let $\delta(a,b)$ denote the distance between two points. And let A and B be subsets of a metric space V. Then the distance between points is easily extended to a distance between a point and a subset as follows:

$$\delta(a, B) = \inf_{b \in B} \delta(a,b).$$

In this formula, inf denotes the largest lower bound. Loosely speaking, this is the minimum point and is more general in the sense that it need not lie inside the set. As a simple example, the infinite set

$$\left\{ \frac{1}{n} \text{ for } n \in \mathbb{N} \right\}$$

has no minimum. The numbers get smaller and smaller; hence, none of the elements is the smallest element. However, there are lots of lower bounds to this set: all negative numbers and zero. The largest of these lower bounds is given by zero. Hence, the infimum is given by zero in this case. In cases where the minimum exists, the smallest element also provides the largest lower bound and, hence, the infimum equals the minimum. In this way, the infimum is a generalized minimum. The complete construction can be mirrored for maxima. The smallest upper bound of a set is called supremum.

Now that we are able to calculate the distance of a point and a set by using the distances of all points, it is reasonable to proceed along the same lines and vary the first parameter to obtain a distance between two sets. In the first case, it was reasonable to take the shortest distance. However, in this case, we have to use the largest distance instead:

$$\delta(A, B) = \sup_{a \in A} \delta(a, B) = \sup_{a \in A} \inf_{b \in B} \delta(a,b).$$

If we took the smallest distance, a single point in both sets $p \in A$ and $p \in B$ would imply that the distance is zero. The definition above, however, is quite reasonable: for a set A, we have $\delta(A, A) = 0$, it can be seen that the triangle inequality holds as well. However, $\delta(A, B)$ is neither symmetric nor does zero distance imply equality. However, $\delta(A, B) = 0$ implies $A \subseteq B$. This is quite easy to see: as all numbers are nonnegative, $\delta(A, B) = 0$ implies $\delta(a, B) = 0$ for all $a \in A$. This means that for all $a \in A$, we can find $b \in B$ such that $\delta(a,b) = 0$. This, however, implies that $a = b$. Together, we find all $a \in A$ inside the set B. Consequently, $A \subseteq B$. A simple trick now completes the Hausdorff distance. Symmetry is constructed by taking the maximum of $\delta(A, B)$ and $\delta(B, A)$.

Fig. 6.2 Example of small
Hausdorff distance for
different trajectories

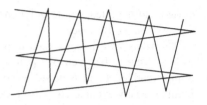

For the purpose of comparing trajectories given by functions A and B, we denote
the image of the functions as $I(A)$ and $I(B)$, respectively. For trajectories A and B,
their Hausdorff distance is then finally given by

$$\delta_{HD}(A, B) = \max \begin{cases} \sup_{a \in I(A)} \inf_{b \in I(B)} \delta(a, b) \\ \sup_{b \in I(B)} \inf_{a \in I(A)} \delta(a, b). \end{cases}$$

In order to calculate this distance function, it is customary to use a finite
polygonal curve approximation to the trajectory, which is fine enough such that the
discretization error is acceptably bounded. For such polygonal tracks, it is possible
to calculate the Hausdorff distance exactly by analyzing a finite number of critical
values.

Though the Hausdorff distance can be efficiently computed for polygonal curves
and is strongly rooted in theory, it has difficulties in differentiating between non-
similar trajectories in some cases. A very typical example is given in Fig. 6.2 in
which the two depicted curves have a comparably small Hausdorff distance.

This small Hausdorff distance is returned due to the Hausdorff distance ignoring
the partial ordering of a trajectory. For these cases, another classical metric has been
proposed by Fréchet.

6.3.2   Fréchet Distance

The Fréchet distance can be thought of as a distance, which shall compare the image
of the trajectory functions independent of the actual time, but still dependent on the
ordering, such that the same trajectory with different speeds has zero distance while
two different trajectories should never have zero distance.

The main problem of trajectory distance computation is to reduce the numerous
differences between different points inside the location into a single number. Fréchet
proposed a simple geometric idea for that, which is best described in an informal
way as follows.

Imagine a dog and his owner moving on two separate paths. They are allowed
to move in their own respective speed, but not to move backwards. The Fréchet
distance is given as the minimum length of a leash needed to keep the dog and its
owner connected for all allowed movements.

Formalizing this idea, the movement rules can be given by allowing all reparametrizations of the trajectory functions. A reparametrization in this context is a continuous, non-decreasing, surjective map of the unit interval into itself. Minimizing over all possible reparametrizations yields the minimum distance for all possible movements of the owner and the dog in the informal definition. For each reparametrizations, the candidate Fréchet distance is given by the length of the leash needed for the actual reparametrization, which is given by the largest distance in which the owner and the dog are during the reparametrized trajectory leading to an overall definition of the Fréchet distance of two trajectories A and B as follows:

$$\delta_F(A, B) = \inf_{\alpha,\beta} \sup_{t \in [0,1]} \delta(A(\alpha(t)), B(\beta(t))),$$

where $\alpha, \beta : [0, 1] \rightarrow [0, 1]$ are reparametrizations.

By minimizing over all possible reparametrizations, the sequential structure of the trajectory is the central ingredient to the distance. For example, the trajectory in Fig. 6.2 has small Hausdorff distance. However, it has large Fréchet distance.

Though this definition seems to be intractable for a computer implementation as the infimum over all functions of a very general class of functions is usually impossible to compute in finite time, it has been shown that in the special case of polygonal curves, efficient calculation becomes possible.

Let p and q denote the number of vertices in the tracks A and B. Then a fast algorithm to calculate the Fréchet distance is given by Alt and Godau in [1] and has a running time of $\mathcal{O}((p^2q + q^2p) \log pq)$. In many cases, the complex discussion of critical points on the interior of polygonal line segments is not too important. With an error bound by the length of the longest segment in the polygonal curves, it suffices to compute the Fréchet distance only at the endpoints of segments for a running time of $\mathcal{O}(pq)$.

6.3.3 Jaccard Distance

A fundamentally different way to reduce the complexity of trajectories is given by fixing a cell subdivision of the area and measuring similarity as the set similarity between the sets of cells, which the trajectory enters. These cells can be given by a cellular network such as GSM or Wi-Fi, by some space division algorithms including kD-trees and others, or being induced from semantic building models as room names or room functionalities. Overall, the spatial nature of a trajectory is transformed into a purely combinatorial structure of a sequence of letters over an alphabet. The letters are given by the space subdivision, and the alphabet is given by all possible places.

Though there are a lot of possible distances for such sequences of letters, one of the most important is given by the Jaccard distance.

The *Jaccard index* is quite simple and compares size of the intersection of two sets with the size of the union of two sets as follows:

$$J(A, B) = \frac{|A| \cap |B|}{|A \cup B|},$$

where A and B are considered as sets and, hence, the ordering of the symbols in the trajectory is completely ignored. This Jaccard index takes values between $0 \le J(A, B) \le 1$. A Jaccard index of 0 implies an empty intersection, that is, two completely different subsets. A Jaccard index of 1 implies equality. Turning this over, we can define a measure that actually is a metric by setting

$$\delta_J(A, B) = 1 - J(A, B).$$

This can readily be used to compare two sets, and using some cell division of the spatial domain, this can be applied to trajectories as well. Though the Jaccard metric removes a lot of information and shares the same problem as the Hausdorff distance, the computational approximation of the Jaccard distance is possible without a tedious pairwise comparison of all trajectories with each other. Therefore, a technique called locality-sensitive hashing can be used. Locality-sensitive hashing is a method of hashing, where the numerical outcome of the hash function is equal for similar trajectories and different for others. Then, similar trajectories can be retrieved by comparing the integer numbers assigned to the trajectories. In this case, efficient data structures for retrieving the objects with the same hash from a database can be applied.

A simple scheme for providing locality-sensitive hashing is given by the following construction. Let h be a uniform, quasi-random hash function mapping a cell id to an integer $\{0, 1, \ldots, N - 1\}$, where N denotes the number of cells in the cell subdivision. Let further A and B be the sets of cells, which are hit by two trajectories. For a set, we summarize the individual values of the chosen hash function h defining $H(A)$ as follows:

$$H(A) = \min_{a \in A} h(a).$$

This function has the welcome property that it probabilistically recovers the Jaccard index:

$$P(H(A) == H(B)) = J(A, B).$$

Using the value of H directly for the bucket leaves us with a lot of false positives. A trick can be applied to mitigate this problem: using not only the smallest element for defining the bucket, but also the second smallest, reduces the number of false positives drastically. That is, a hash signature $\tilde{H}(A)$ is defined by

$$\tilde{H}(A) = (H(A), H(A \setminus \{c\}),$$

where c is the element of A, which leads to the minimum in $H(A)$. In other words, c is the element with $H(A) = h(c)$. In order to reduce the number of false negatives, this procedure can be iterated using different hash functions. Similarity can then be defined as having the same hash signature $\tilde{H}$ with respect to a specific number of different input hash functions h.

The previously described distance estimates for trajectories have strong mathematical backings. They fulfill the properties of a metric, which amounts to the properties one should expect for a reasonable distance. This, however, leads to relatively complicated distances in cases of Hausdorff and Fréchet or minimalistic distances in the case of the Jaccard metric, which depends on the cell subdivision and does not take the sequential structure of a trajectory into account. In order to assess some sort of similarity measure to trajectories, one can drop the aim for a complete metric and design different simple functions, which can be used to assess some aspects of trajectory similarity.

6.3.4 Closet Pair Distance

A minimalistic example of such a distance is given by the closest pair distance. Instead of studying the entire trajectories and comparing them, we can compare a pair of corresponding points in the two trajectories with respect to some property. For the closest pair distance, from both trajectories the points are chosen, which have the overall minimal Euclidean distance, and the trajectory similarity is measured as the distance of these two nearest points:

$$\delta_{\text{CPD}} = \min_{a_i \in A, b_i \in B} d(a_i, b_i).$$

Obviously, very different trajectories can have a small closet pair distance (CPD). However, this distance can be used as an efficient filter in many cases, because very similar trajectories cannot have large CPD. Thus, this distance cannot be used for a detailed comparison of trajectories; however, it can prune trajectories from more complicated computations, which have no chance of being near to each other.

The idea of using some definition of corresponding points can be extended to several constructions, which try to integrate more information into the similarity measure. A prominent example is given by the Euclidean distance sum (EDS).

6.3.5 Euclidean Distance Sum

A very simple way to extend the Euclidean distance for the comparison of trajectories is given by fixing a correspondence of points in both trajectories and summing up the distances between pairs of corresponding points. In simple cases, this correspondence is given by equality of timestamps: when both trajectories are of the same length and sampled at the same time points, this provides a reasonable correspondence. In this case, the distance of sampled trajectories $A = (a_i)$ and $B = (b_i)$ can be computed as

$$\delta_{\text{EDS}} = \sum_{i=1}^{n} d(a_i, b_i).$$

Unfortunately, this distance is limited to trajectories of equal length and equal samplings. Moreover, if the movement speed of the objects in both trajectories is different, then this distance will be nonzero even for identical trajectories. To mitigate these problems, an alternative definition of correspondence has been given and results in the dynamic time warping (DTW) distance.

6.3.6   Dynamic Time Warping

In order to overcome the need of equal length and equally sampled trajectories, a heuristic has been proposed to find a reasonable set of correspondences. The DTW distance is similar to the EDS distance in that it is defined as the sum over distances between corresponding points. However, the correspondence need not be one to one, and for DTW, it is allowed to use the same point in one trajectory more than once. Still, all points have to be considered at least once and used in the correct ordering. Concretely, this leads to three choices for each step of calculation. This definition leads to the following dynamic programming formulation, which can be used to calculate the DTW distance:

$$\delta_{\text{DTW}}(a_{1...n}, b_{1...m}) = d(a_n, b_m) + \min \begin{cases} \delta_{\text{DTW}}(a_{1...n-1}, b_{1...m-1}) \\ \delta_{\text{DTW}}(a_{1...n-1}, b_{1...m}) \\ \delta_{\text{DTW}}(a_{1...n}, b_{1...m-1}). \end{cases}$$

This distance can be computed in $\mathcal{O}(mn)$ time [2]. Note that this definition does not provide a metric as the triangle inequality is not always fulfilled. However, this distance definition has found wide adoption in the domain of trajectory computing for its simplicity, efficiency, and practical discriminative power.

6.3.7   Longest Common Subsequence (LCSS)

While the previous definitions capture a lot of different characteristics of trajectories and can be used to compare them, they are all sensitive to noise and outliers. In all previous definitions, all points have had to be used for matching; hence, outliers always contribute to the distance function. In order to flexibly treat short deviations in trajectories as well as measurement outliers, the previous definition has motivated a new paradigm in which points can be occasionally left out. However, this freedom has to be controlled. Otherwise, all points could be left out for an all-zero distance function. Therefore, the number δ controls how many measurements can be left out in order to get a better matching, and the number ϵ ensures that important points cannot be left unmatched

$$
\delta_{\text{LCSS}}(a_{1...n}, b_{1...m}) = \begin{cases} 0, \text{if } n = 0 \text{ or } m = 0 \\ 1 + \delta_{\text{LCSS}}(a_{2...n}, b_{2...m}) \text{ if } d(a_1, a_2) \leq \epsilon \text{ and } |m - n| < \delta \\ \max\left(\delta_{\text{LCSS}}(a_{2...n}, b_{1...m}), \delta_{\text{LCSS}}(a_{1...n}, b_{2...m})\right). \end{cases}
$$

This definition is essentially summing up the value of one for each correspondence and allows some time adjustment. Sometimes it is useful to normalize this definition of distance with respect to the number of samples used inside the definition. Therefore, it has been proposed to divide the result by the number of samples in the trajectory with fewer samples:

$$
\tilde{\delta}_{\text{LCSS}}(a_{1...n}, b_{1...m}) = \frac{\delta_{\text{LCSS}}(a_{1...n}, b_{1...m})}{\min(n, m)}.
$$

6.3.8 Edit Distance on Real Sequences

Longest common subsequence already uses a parameter ϵ and only binary differentiates between matching points and non-matching points. The actual difference $d(a, b)$ is not added to the measure. Instead the number of matching points is counted. With this strategy in mind, it is natural to think of similarity from an amount of work discussion. How many edit operations are needed to transform the sequence of matching points into each other? For edit distance on real sequences (EDR) distance, the edit operations are insert, delete, and replace and the EDR distance returns the number of edit operations needed to change A into B.

The EDR distance can be given by the following recursive process:

$$
\delta_{\text{EDR}}(a_{1...n}, b_{1...m}) = \begin{cases} n, \text{if } m = 0 \\ m, \text{if } n = 0 \\ \min \begin{cases} \delta_{\text{EDR}}(a_{2...n}, b_{2...m}) + \begin{cases} 0, \text{if } d(a_1, b_1) \leq \epsilon \\ 1, \text{else} \end{cases} \\ \delta_{\text{EDR}}(a_{2...n}, b_{1...m}) + 1 \\ \delta_{\text{EDR}}(a_{1...n}, b_{2...m}) + 1. \end{cases} \end{cases}
$$

The minimum is taken, whenever both trajectories contain points. Replacing the distance $d(a_1, b_1)$ by either zero or one makes this distance handle noise. This distance has proven useful in practical considerations. Unfortunately, it lacks the triangle inequality. Of course, it is of general interest whether distance functions exist, which can support time shifting such as DTW, LCSS, or EDR as well as being metric such as the Euclidean, Hausdorff, and Fréchet distances. A beautiful trick allows for that leading to the Edit Distance with Real Penalties (EDR).

6.3.9   Edit Distance with Real Penalties

By a detailed analysis of why EDR and DTW do not fulfill the triangle inequality, a trick is used in order to correct for the problems. While DTW introduces problems by repeating points, EDR introduces problems by using constant cost for all operations. The following definition, which is quite similar to EDR, fulfills the triangle inequality by overcoming both problems. Therefore, a constant helping point g is introduced, which can be set to zero as proposed by the inventors of edit distance with real penalties (ERP) [3]:

$$\delta_{\mathrm{ERP}}(a_{1...n}, b_{1...m}) = \begin{cases} \sum_{i=1}^{n} |a_i - g|, \text{ if } m = 0 \\ \sum_{i=1}^{m} |b_i - g|, \text{ if } n = 0 \\ \min \begin{cases} \delta_{\mathrm{ERP}}(a_{2...n}, b_{2...m}) + d(a_1, b_1) \\ \delta_{\mathrm{ERP}}(a_{2...n}, b_{1...m}) + d(a_1, g) \\ \delta_{\mathrm{ERP}}(a_{1...n}, b_{2...m}) + d(b_1, g). \end{cases} \end{cases}$$

The choice of the point g, of course, changes the value of ERP. However, trajectory distances are usually used to compare different trajectories, and if the trajectories are equal, then the point g is never used. ERP provides a proper metric and still encodes the ideas of edit distance.

6.3.10   Outlook

The nature of trajectories is complicated and this is reflected by the numerous different distance definitions and similarity scores. The previously discussed distance functions have been chosen, because they are either well rooted in theory or have motivated a lot of research. In general, three approaches have been proposed: firstly, extensive and simple extension of the existing metric to trajectories; secondly, transforming the trajectory into a sequence over a finite alphabet; and thirdly, a marriage of both approaches. New distance definitions are proposed regularly and the aim of measuring specific different aspects of trajectories for different applications makes this field beautiful and worth exploring.

6.4   Trajectory Computing for Indoor LBS

The area of trajectory computing is defined in a rather general way. With respect to indoor location-based services, several aspects of trajectory computing become more important than others. This section will provide some examples of applying results from this newly developing field to indoor location-based services.

The main motivation of considering trajectories instead of single location fixes is given by the assumption that the measurement and location errors are uncorrelated over time. Hence, a trajectory consisting of the current position and several previous positions contains more information about the current location than the last position fix alone. Moreover, location-based services are often human-centered services and rely on personalization. From a trajectory, the situation of a user is better extractable than from a single location. In this way, more complex situations can be understood and differentiated for better personalization results.

6.4.1   Trajectory Computing for Positioning

For positioning applications, trajectory computing can be applied in two different ways: either it can be used to infer a better position out of a history of positions just as the techniques of Chap. 5 or it can be used to infer first similar trajectories from a dataset and reason about properties of these trajectories (e.g., the next point of interest being visited). This modified positioning problem is, however, more difficult to evaluate as there are two sources of problems: firstly, a sufficiently similar trajectory to the current trajectory has to be selected from the database, which might be impossible, if no such trajectory exists, and, secondly, the time inside this trajectory has to be estimated. The most important strength of this approach is that it can exploit reproducible errors for increasing discriminativity instead of filtering these unwanted errors away.

For a concrete example, consider a signal-strength-based localization approach employing lateration. Figure 6.3 depicts a situation, where the user is passing by a hallway containing an access point. The dashed line represents the true trajectory of the user. A lateration-based positioning approach will suddenly attract the sequence of position fixes towards the location of the access point, once a free line-of-sight transmission becomes possible. In a trajectory computing approach, this sudden increase in signal strength is modeled already in the database and is to be expected. Hence, the trajectory-based approach does not react onto this falsely detected trend.

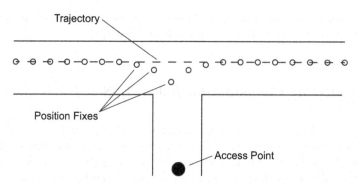

Fig. 6.3 Suddenly changing signal strength does affect position fixes, while it leaves trajectory fixes untouched

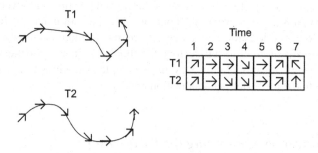

Fig. 6.4 Extracting REMO matrices from trajectories

6.4.2 Movement Patterns

A classical approach to trajectory-based services is given by considering the relative motion of different individuals with respect to each other. The REMO framework (relative motion) is a simple example of how this can be done: it transforms raw trajectory data into sequences of discrete motion directions (often eight directions) called azimuth. A trajectory is therefore represented as a vector containing these discrete directions. A trajectory dataset is then naturally represented as a matrix, and this matrix is used to infer about relationships of trajectories. Figure 6.4 depicts this mechanism. This process greatly simplifies trajectories and hence their relationships. This simplification leads to simple definitions for the following three basic movement patterns.

6.4.2.1 Basic Movement Patterns

The movement pattern of *constance* captures intervals of constant azimuth such as 2, 3 in the first trajectory or 3, 4 in the second trajectory. While this movement pattern captures the same value along the horizontal axis of a REMO matrix, the

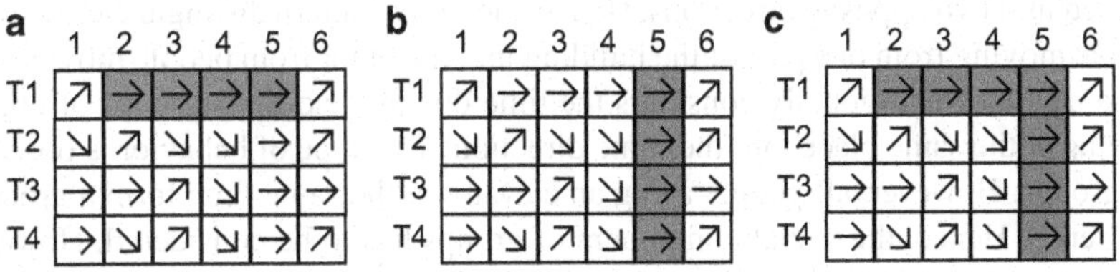

Fig. 6.5 Basic movement patterns in a REMO matrix. (**a**) Constance. (**b**) Concurrence. (**c**) Trendsetter

movement pattern of *concurrence* captures same values along the vertical axis: multiple objects move at the same time into the same direction. This is given in the example at times $1, 2, 4, 5, 6$. The final basic movement pattern is given as a combination of these two patterns: the *trendsetter* is the first occurrence of a *constance* pattern such that after some time, other trajectories join the same movement resulting in a *concurrence* pattern. Figure 6.5 depicts REMO these three basic patterns inside a REMO matrix.

In this approach, trajectories are represented as a finite vector of discrete values capturing only coarse time information and no location information at all. Therefore, in practice, too many irrelevant patterns would be detected, and hence, the search space has to be restricted to meaningful patterns with some temporal or spatial nearness to each other.

6.4.3 Spatial Movement Patterns

Therefore, these basic movement patterns are combined with spatial constraints. The original REMO concept supports spatial constraints in a very general manner and names a few examples of measures capturing different spatial nearness. For the ease of exposition, we will restrict the exposition to the simple case of limiting to a spatial circle, while any other spatial constraint can be integrated in the same way. Given a spatial constraint, we can derive three new patterns from the three basic patterns as follows: the movement pattern *Track* is defined to be a movement pattern of constancy inside a spatially constrained region. The movement pattern of a *Flock* is defined to be a pattern occurrence of *Concurrence* inside a spatially constrained region. The pattern of *Leadership* is defined to be a spatially constrained pattern of *Trendsetter*.

These combined motion patterns are best explained with simple examples. The *Track* movement pattern can be observed for people using an escalator in buildings. The escalator defines the direction of movement, and if the spatial constraint is chosen appropriately to contain only the escalator, this leads to an observed constancy pattern inside the spatial region. There are other topological features of buildings, which can be observed as *Track* patterns. For example, hallways tend to be used into two opposite directions. Hence, the *Track* pattern distinguishes between people moving from one part of the building into the other from people moving vice versa. A *Flock* additionally considers the time domain and finds groups of objects moving at the same place into the same direction. This type of behavior is observed for groups. However, all people using an elevator at the same time form a group in this sense. Hence, the social component of groups has to be considered. However, if a group of moving objects is observed in building a flock at different times and locations, they are very likely belonging to the same social group. This information can already be used for location-based services in which a false positive is not harmful. For example, users building a flock can be given a coupon for a reasonable location in the expected future location. If some user is informed about this coupon, but can actually not use it, this is not a great problem. The *Leadership* movement pattern detects situations in which a group is actually starting a joint activity.

6.4.4   Group-Based Motion Patterns

Additional to the movement patterns motivated from the REMO matrix approach together with spatial constraints, several other patterns have been defined and studied, which can be used to extract meaningful information out of a database of trajectories. These patterns cover—in addition to joint movement—dynamic spatial relations between different objects.

The movement pattern of *Convergence* is defined to be a group of mobile entities, whose extrapolated motion intersects inside a limited spatial region. For extrapolating the motion, the current motion azimuth can be used to define a line or a navigation graph can be utilized. If personalized trajectory histories are available, the most probable personal trajectory can be used as an extrapolation of the current trajectory fragment. The *Convergence* pattern is useful for location-based services as it can predict locations, where a user can deposit something for another user, who will be able to pick it up without additional effort. Orthogonal to the *Convergence* pattern, a pattern of *Divergence* is defined. The *Divergence* pattern is given, when a set of users heads for leaving a limited spatial region in mutually different directions. The patterns of *Convergence* and *Divergence* can also be combined with a temporal constraint resulting in movement patterns *Encounter* and *Breakup*. They detect meetings when the current motion is extrapolated into the future. In addition to the patterns of *Convergence* and *Divergence*, these patterns require that the extrapolated location at some time interval intersect in a limited spatial region. These two movement patterns can be used to assist users in planning future meetings by informing users where and when they can actually meet with minimal overhead. Other patterns have been defined for detecting groups of objects traveling together. There are several approaches based on clustering in which several objects are a group if they form a cluster for a given interval of time. One example of such a pattern is given by the *Convoy* pattern.

This simple approach of clustering is, however, a bit difficult to use in practice, as short deviations from the route or short times of stopping could break the cluster condition. Therefore, a relaxed pattern called *Swarm* has been defined. A *Swarm* is a set of at least m objects contained in the same cluster for at least n timestamps. However, these timestamps need not be consecutive. In this way, the precise geometry of the trajectory becomes unimportant as long as the cluster condition is not broken. Thereby, the impact of different movement dynamics as well as measurement noise is reduced. This can be quite important inside buildings: at an airport security check, for example, individuals of a group often take a different gate in parallel. Due to the high density of people waiting at a security check, it is very likely that clusters will collect people at the same security gate instead of people actually moving together. These problematic places can be left out automatically by only specifying a minimal amount of time of joint movement.

6.5 Summary

Trajectory computing is a hot topic and still in its infancy. Though some useful movement patterns have been defined, the author expects trajectory computing to reveal a lot of social information to computer systems. The challenge mainly lies in dealing with the uncertainty of conclusions drawn from noisy trajectories and in finding a balance between the poles of usefulness and privacy. The services offered on the basis of trajectory data must be constructed in a way which respects the individual privacy while still being useful for the user and his peers most of the time. Moreover, services based on trajectory computing have a higher risk of interrupting a user with useless information due to the complexity of situations being predicted. For example, a system might perfectly detect a *Convergence* pattern at a meeting place. However, it is needless to inform anyone about this observed pattern, as there is a meeting planned at that specific location. The question whether observing a complex movement pattern is an interesting novel insight for the user or a result from his current intention is difficult to answer. This poses a challenge on designing a non-interrupting human–computer interface. In general, the integration of context information and location information provides a great chance, but also a great challenge.

6.6 Further Reading

The area of trajectory computing is a very active research area. While it is based on very classic results including Fréchet distance and Hausdorff distance, many substantial contributions have been made only recently. A good selection of articles can be found in the book *Computing with Spatial Trajectories* edited by Zheng and Zhou [7]. Furthermore, many classical and up-to-date results are being published in the *IEEE Transactions on Knowledge and Data Engineering*. For the topic of relative motion patterns, the original paper is very informative [6]. For extensions towards more flexible treatment of moving objects, the theory of density connectedness has raised much attention [5] as it is able to overcome thresholding problems in datasets with noise by defining nearness of spatial objects in a more flexible manner. A taxonomy of movement patterns has been proposed by Dodge and Weibel, which can serve as an additional overview to the topic [4].

References

1. Alt, H., Godau, M.: Computing the fréchet distance between two polygonal curves. Int. J. Comput. Geom. Appl. **5**(1), 75–91 (1995)
2. Berndt, D.J., Clifford, J.: Using dynamic time warping to find patterns in time series. In: KDD Workshop, Seattle, WA, vol. 10, pp. 359–370 (1994)
3. Chen, L., Ng, R.: On the marriage of lp-norms and edit distance. In: Proceedings of the Thirtieth International Conference on Very Large Data Bases, pp. 792–803 (2004)
4. Dodge, S., Weibel, R., Lautenschütz, A.K.: Towards a taxonomy of movement patterns. Inf. Vis. **7**(3–4), 240–252 (2008)
5. Ester, M., Kriegel, H.P., Sander, J., Xu, X.: A density-based algorithm for discovering clusters in large spatial databases with noise. In: Kdd, vol. 96, pp. 226–231 (1996)
6. Laube, P., Imfeld, S., Weibel, R.: Discovering relative motion patterns in groups of moving point objects. Int. J. Geogr. Inf. Sci. **19**(6), 639–668 (2005)
7. Zheng, Y., Zhou, X.: Computing with Spatial Trajectories. Springer, Berlin (2011)

Chapter 7
Event Detection for Indoor LBS

Never say "I was wrong". Instead, say "What an interesting turn of events"!

Anonymous

Location-based services are constructed from observing the surroundings and the situation of a mobile device. While very basic location-based services outside are only observing a single sensor, the GPS chip, the inherent complexities of buildings and the localization problem inside buildings make most indoor location-based services use a multitude of environmental information from a lot of different sources, possibly including the building itself. While many of these sources are available in many areas, some might be only partially available or randomly available. This includes, of course, GPS satellite sightings, which can occur inside buildings, as well as Wi-Fi beacons, which can be detected or not. Furthermore, a building can provide an interface for providing information to a mobile device or it does not. Furthermore, dedicated beacon technology such as iBeacon has been presented in order to provide proximity awareness to smartphone applications. In order to manage these situations in which a lot of sources of information are to be integrated, it is customary to define the software infrastructure using an event-driven architecture.

As a preliminary example, think of a turn-by-turn navigation system. For such a system, the actual location, possibly given by GPS outside or by some indoor localization system inside, is not that much of interest. It often suffices to detect the location of the user with respect to the planned route: where inside the route is the user currently or did the user leave the planned route. In this setting, the optimal stream of navigational events usually consists only of changes of movement direction. The basic semantic is in most cases to follow along the route until the next instruction tells something different. Thereby, a lot of junctions and decisions need not be explained to the user, and the cognitive load induced by navigation keeps low.

This navigational experience is what people are missing inside buildings. Basically, one is interested in a sequence of instructions which enables users to find the best way and which is short. For people movement, however, the structure of the navigational space is not as limited as for vehicle navigation outside buildings. While a common car navigation system only needs a very limited set of instructions, which are easy to generate from map information and a shortest path, the same type

© Springer International Publishing Switzerland 2014
M. Werner, *Indoor Location-Based Services*, DOI 10.1007/978-3-319-10699-1_7

of instruction is useless inside buildings. The previous chapters have shown that we are able to find the actual position of a moving object inside buildings with limited quality and that there are a lot of techniques to deal with the inherent limitations and complexities of indoor navigation and mapping. However, we have not concentrated on the problem of guidance, namely, how to explain the shortest path from the currently inferred position for a mobile user such that this user is able to follow the path.

For a complete and successful navigation system, the complete system must be useful for the user in that it allows him to reach a given destination easier than without the system. This task of guidance, guiding a user towards a goal, can be subdivided into four subtasks following [1]:

- **Orientation**: The system must have enough information to find his own position.
- **Choice of Route**: The system must have enough information to find and choose a route towards a target. This could be a shortest or otherwise preferred way.
- **Following This Route**: The system must provide the user with enough information to actually follow this route. This includes that the user is always aware of whether he is still on the correct route and what the next navigational decisions are.
- **Detection of the Target**: The system must provide the user with sufficient information to detect that he actually reached the target location and that the guidance process will stop.

The previous chapters have been limited to providing technological aids for these tasks only: *orientation* can be provided using positioning techniques developed in Chap. 3. Moreover, sufficient environmental models (Chap. 4) provide the system with awareness of points of interest and possible routes and can be used to provide the *choice of route*. Having this technical description of a route, for example, as a series of line segments, we can again use positioning techniques or trajectory matching techniques to continuously check that the mobile target resides on this route providing *Following the Route* and, finally, we are able to infer the situation where the user has reached a target through positioning.

In summary, the techniques from the previous chapters provide us with a complete navigation system. However, the user has been left out of consideration. The question for this chapter is how this technical system can be extended to include the user. We have now enabled a system (e.g., a smartphone) with abilities for the complete guidance process, but we have not yet explained how this can be helpful for a user.

For indoor navigation, it is commonly the case that the system has only limited communication capabilities towards the user. Either the system can only communicate from time to time (e.g., when the user stands and looks at his smartphone or when the user is located in front of a distributed stationary display system) or via a limited channel (e.g., only audio or even only vibrations). And even if the scenario includes augmented reality and has unlimited communication capabilities to the user, the system should still provide service at minimum cognitive load to the user. The need of reducing cognitive load can also be motivated from car navigation systems: though most car navigation systems provide audiovisual guidance, the audio instructions are in general more useful than the visual aids as the visual cognition is used by the driver for its primary task. Hence, using visual aids distracts the user and increases risk, while audio guidance does not distract from the primary task of driving.

In all these cases, the communication channel between the system and the user should provide enough information to actually achieve the goal, but not too much information. For indoor navigation, it is a very complicated question, how to provide enough information for a mobile user. Human beings are used to infer their orientation from their surroundings and their previous experience. Hence, in guiding humans, it is of central interest to provide information enabling the user to orient in space. Therefore, humans are mainly using landmarks, which are characteristic appearances along a route. To understand the use of landmarks for the orientation task, compare the following two instructions:

- Go straight ahead for 100 m and then turn right. Go further ahead for 50 m and you reach your goal.
- Go straight ahead. In front of the staircase, turn right. At the end of the hallway, you find your goal.

The first instruction is complete and correct; however, it is difficult for the user to follow along the instruction as he might be unable to correctly estimate when he has moved 100 m. For the second case, the user can simply memorize the three words "staircase," "right," and "end" to easily complete the *Orientation* task and partly the *Following a Route* task.

Landmarks are descriptions of objects inside the navigational space, which are locally distinctive. Typical landmarks in buildings include staircases, doors, corners, artwork, classical signs, room numbers, etc. These landmarks can be classified into three categories with respect to their possible use for a guidance task:

- **Confirming Landmarks:** These are all types of landmarks that are easily visible from the correct route.
- **Segmenting Landmarks:** Landmarks that can explain the change of instructions from the instruction stream.
- **Disconfirming Landmarks:** Landmarks, which inform the user that he has left the planned route.

Confirming landmarks can be given to confirm the user that he is still following the correct route and thus enable the user to perform the *Following a Route* task. Therefore, the landmarks should be distinctive with respect to wrong alternatives. In this class, a lot of landmarks can be chosen, which are clearly not distinctive themselves (e.g., a staircase) but which provide confirmation due to their sequence of occurrence.

Segmenting landmarks are landmarks which can be used to activate a new instruction. These are highly important for the user as this user will often have problems estimating distances and hence finding the segmentation points between instructions. An illustrative example is given by the instruction "After the third staircases turn right."

Disconfirming landmarks are landmarks which might or might not be visible from the planned route but which reside on probable wrong ways and inform the user that he has left the planned route. An example of using disconfirming landmarks in explanations is given by: "Simply go ahead. Then at the corner before the street turns left, turn right. When you see a fountain, you have missed the turn." This example shows the usefulness of disconfirming landmarks. Sometimes it is too complex to describe the correct decision. There could, for example, be no distinctive landmarks near the correct turn. However, following the general advice "go ahead," the user might find himself at a highly distinctive landmark off the optimal route from which he can easily turn back to the planned route.

Landmarks give a motivation for the general concept of this chapter. From a user perspective, navigation does not only consist of a continuous stream of information for the user but can greatly be supported by a multitude of different events. Combining the examples above, we can explain the navigation system as a sequence of events including important turns as well as useful side information to confirm the user that he is still on the right way or to inform the user that he should reorient, possibly stop and look at his smartphone for further instructions.

7.1   Event-Driven Applications

An event-driven application is an application for which the complete functionality consists of detecting and reacting to events. In these areas, an event is defined to be a significant change of some situation. In reaction to an event, an event handler is usually invoked, which defines the reaction to the event. These event handlers can also generate new events and in this way trigger other event handlers. It is also common sense that event handlers can access the system state and the list of expected or planned as well as the list of previous events. This concept has been widely applied to message-based middleware solutions, which can lead to responsive and efficient systems. Event-driven applications consist of several core elements, which can be arranged in a layered infrastructure as depicted in Fig. 7.1.

Event processing can be split into two different architectures:

In *simple event processing*, an event occurs and triggers an event handler, which takes some action solely based on the event. A precondition of simple event processing is a very good model of events, as the modeling of events has a severe influence on the achievable functionality. In simple event processing, an event triggers the same event handler independent from different events and states, and hence, when the granularity is too fine, the event handler might not have enough information to react reasonably, or, when the granularity is too coarse, the event handler might get invoked in pretty different situations and cannot react uniformly.

Fig. 7.1 Layers of
event-driven applications

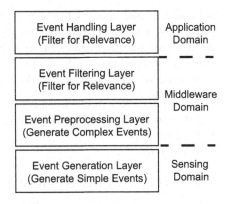

In *complex event processing*, pattern detection and computational intelligence are used to infer so-called complex events from multiple occurring simple events. The interrelations between simple events and their relation to complex events can be arbitrary, and hence, complex event processing has a wider applicability as it completely decouples naturally occurring simple events (e.g., measurements) from application-centered useful events (e.g., events that change the situation of a user). A special form of complex event processing is given by *event stream processing*, where the time of occurrence of events is a central element in the generation of complex events. For many applications, however, both approaches lead to very similar systems as the relevance of a simple event is often highly correlated to the current time.

In order to flexibly integrate various sensors and sources of information into one system, using event-driven architectures for indoor location-based services is a quite natural approach.

From one perspective, it is the case that most measurements characterizing the situation of the mobile entity can be formulated as events at measurement time and that a lot of these measurements take place in a distributed, non-cooperating infrastructure. A proximity-based indoor positioning system, for example, can provide events for users entering a region or leaving a region, but cannot infer on the relevance of this information, as they have no information about the current task of the mobile entity.

7.2   Event Sources for Indoor Navigation

The first layer in event-driven applications is of course to describe where primary and simple events come from and how they look like. Though this does not fit into the layered structure of event-driven applications, it can be useful to have an application-oriented motivation for constructing simple events. It is worth noting that most of the events will occur during ongoing navigation, hence need to support the Following the Route and Detection of the Target subprocess of the navigational process. We will organize this non-exhaustive list of events by their technical source. In this way, the discussion can be easily extended with new sensors and new technologies coming up in the future.

Possible sources for events in indoor location-based services are:

- Knowledge about the environment, environmental models
- Infrastructure and building automation
- User interface
- Positioning
- Activity recognition
- Other context information (personal profiles, time, calendar, etc.)

For this discussion, an event will consist of a point in time (either future, current, or past) and be linked to a set of information describing the event. Hence, all events constitute a temporal stream of events, and inside this stream, the current time will be used to differentiate between past, current, and future events. Note that future events are often called planned events. However, this is not clear enough, as this would mean that events in the future have all been planned by someone or something. Sometimes, however, a system knows that some event will happen in the future without that it was planned by someone. This introduces a subtlety, which unfortunately is often overlooked in system design: the time associated with an event in the event-driven system must be useful for the application. As a consequence, either the event source must have a synchronized clock (e.g., using GPS or some other variant of distributed time synchronization) or the communication must be real time. Real-time communication has nothing to do with the actual performance of communication and is, hence, not that difficult to achieve, because real-time communication is given if an upper bound to the communication delay can be given. Time-of-arrival positioning using Wi-Fi is a great example of this problem: while the hardware timers of the Wi-Fi chip have sufficient resolution, it is nowadays merely impossible to measure the roundtrip time of frames using commodity Wi-Fi devices. This is due to the fact that the communication delay between the Wi-Fi chipset and the operating system kernel is too dynamic due to other interrupts. Hence, the timestamps taken in the operating system kernel could have enough resolution using specific CPU timer registers, but the time is not useful. The only solution is to introduce a timestamp within the Wi-Fi hardware, which has been discussed and standardized in IEEE 802.11v "Timing Measurement." However, this feature has not yet been rolled out in large scale.

Assuming that it is possible to generate a sequence of events with meaningful timing information, this stream of events will be fed to the location-based service, which can then create overall and smart decision on whether or not to inform the user about the current, previous, or expected events and in which way. In total, this enables a navigation system, for example, to adapt the user interface to the actual situation of the mobile user with respect to his navigational task. Furthermore, this stream of events can be generated and used even for cases, where the system does not know the navigational task. Possibly, the user interface can be used to inform the user proactively about interesting points or the system might guess the current navigational task and provide guidance with respect to this inferred navigational task. The following will collect some examples of events that are useful for providing indoor location-based services.

7.2.1   Primary Events from Environmental Knowledge

In event-driven architectures, a simple event is an event, which can be handled as is without the relation to other events, while a complex event is an event, whose interpretation depends on other events. There is another distinction between events, namely, between primary events and secondary events. *Primary events* are events that are directly observable. *Secondary events* are events that are generated out of patterns of primary events. These terminologies are sometimes intermixed due to the fact that most secondary events are generated from event handlers reacting on complex events. Sometimes, the term complex event is used for secondary events, while primary events are called simple events. Note, however, that complex events can be handled by simple event handling and that simple event handling can handle secondary events.

For many indoor location-based services, the environment will be more or less static and does not provide primary events. When the navigational task, however, interacts with any type of infrastructure, this can lead to events from environmental knowledge.

For example timetable services can provide primary events from environmental knowledge, which could be seen as environmental knowledge or as appearing and disappearing points of interest. These timetable services can be given by transport (e.g., bus, taxi, elevator, etc.) or by opening and operating hours. Common examples are entering a zoo, crossing a public park or cemetery, entering a building, crossing streets, etc.

Another class of primary environmental events can be generated in smart city applications from context-awareness. An airport can, for example, measure the waiting time at a security checkpoint and include the result into the environmental database. This example, however, is difficult to distinguish from secondary events, where actual sensory provides the needed primary events, which would then be primary events in the infrastructure class. In some cases, however, these events come up for the application from an existing GIS infrastructure and are, hence, best seen as primary events from environmental knowledge, as the possible sub-events are not directly accessible by the application.

On a larger timescale, of course, a lot of primary events come from environmental models and spatial knowledge: for example, seasonal services such as a Christmas market can be very relevant to the navigational task in that some area is difficult to cross as it is crowded or that some area is accessible during late evenings, which is usually closed at that time.

However, the most common integration of environmental information is into secondary events based on the relation of several primary events to the environment.

7.2.2 Primary Events from Infrastructure

Infrastructure generates a lot of primary events. Usually, all positioning systems based on infrastructure generate regular events such as receiving a beacon from a Wi-Fi access point, associating to a cell tower in a cellular network, lighting conditions (lights being switched on or off), choice of direction of transportation for variable escalators, proximity events, or any other events from distributed sensory.

This type of events is, thus, the most common type. Deploying any sensor system as an infrastructure leads to events, which can be relevant to the application and can be made accessible by an indoor location-based service platform.

7.2.3 Primary Events from User Interface

The user interface is a very special and complicated source of events. The behavior and capabilities of a user interface are quite varying. Consequently, the generation of events from user interfaces is also varying. Still, all cognitive choices which cannot be taken autonomously by the system or for which the user should be notified will generate primary events from user interface. When comparing different setups of user interface and navigation, a lot of different events will enter this class of primary events. As a rule of thumb, one can say that at least all events that are not covered by a different class of events in a specific navigation application will be primary events from user interface. If, for example, the navigation system does not provide a functionality to detect the reach of the target, there will be a mechanism, say, a button, to tell the navigation system that the target has been reached. Another example is given by the case that the navigation system is installed on a smartphone and shall be able to deal with specific disabilities of users, say, not being able to walk stairs. Then there is simply no way to differentiate between disabled users and unimpaired users unless the user interface allows the user to choose.

7.2.4 Primary Events from Positioning

It is customary to also regard positioning events as primary events as there are a lot of solutions, which hide the actual sensor readings from the navigation system. In GPS, for example, it is often impossible to calculate the result of multilateration inside the navigation system software; instead, GPS receivers usually offer an NMEA interface sending over some statistical values (e.g., number of satellites, signal quality measures, time), a velocity estimation, and a position fix location in WGS84 coordinates. Hence, though this system is surely based on deriving these events from sensor events, it makes sense to treat the first event accessible to the navigation platform as a source of primary events. Moreover, a lot of smartphones offer some sort of middleware for location-awareness, which chooses how and from which information to calculate a position more or less autonomously. Again, from the perspective of a location-based service, this type of event is primary, as its predecessor events are inaccessible to the system.

7.2.5 Primary Events from Activity Recognition

Activity recognition is a special form of context awareness in which sensor data is used to classify the current activity of a mobile user from within a set of possible activities. These sets often include only very few and basic differentiations (sitting, walking, standing, walking upstairs, walking downstairs). Fortunately, this classification can be performed easily with common data mining techniques and inertial sensors with sufficient accuracy. Sometimes, this type of event is even classifying between different modes of transportation (e.g., driving a car, cycling, walking, going by train). These primary events are, however, not that relevant to indoor navigation systems. However, they can be very helpful when initiating indoor navigation and in continuous navigation scenarios.

7.2.6 Secondary Events Relevant to the Navigational Task

These primary events are called primary events, because they are either atomic in nature and cannot be split into a sequence or constellation of sub-events or these sub-events are inaccessible to the system. Many of these primary events do not have a useful meaning for the navigational task; however, some of their combinations are quite relevant.

For example, the expected waiting time at a specific escalator is irrelevant to the application unless the user will cross the escalator, and the application needs to estimate the time of arrival or if there is a comparably good route towards the goal avoiding this specific escalator. In these cases, the primary events "Expected waiting time at escalator changed" and "Calculated route contains escalator" combine to a secondary event "Estimated time of arrival for the selected route has changed," which can be very relevant to the system. The system can either compare this selected route to other possible routes or simply inform the user about the changed estimation of arrival by updating the user interface accordingly.

In event-based systems, the calculation of secondary events is often achieved by intelligent agents reacting to events. In most cases, for such an intelligent agent, the events of interest are specified and routed to the agent. The agent usually reacts on incoming events with an event handler and can have a local or global state. Moreover, these agents can create and publish new events, which can be composed from simpler events or from scratch. These secondary events can be put into the current or future event list and can thereby trigger other actions or a callback to the very same agent after some time. Moreover, the agent can change its subscription to events and keep running, if the agent desires to. In this way, it is quite simple to create systems with dynamic behavior and to organize concurrency of task execution. As these systems tend to become complicated and are often distributed in nature, it is customary to select a flexible message-centric middleware as the technical basis for the event-driven system. In this situation, events would be represented by messages, and agents can subscribe to events of interest by some publish-subscribe mechanism provided by the middleware.

7.3 Summary

Having constructed a message-centric middleware, which communicates all events, primary and secondary, towards a set of applications, the applications are left with a multitude of possibly useful reactions to these events. This architecture brings a high amount of flexibility. Furthermore, if several applications can make use of the same event, the event is only generated once. As this can be useful with respect to memory consumption, most mobile operating systems manage sensor and location information in an event-driven way. On the contrary, one drawback of this approach is the complexity of behavior. While event-driven systems are able to behave very flexibly, it is sometimes difficult to understand a concrete behavior and to develop software in this setting.

One open challenge is given by selecting the right trade-off between event generation and user awareness. If too many events are actually communicated through a user interface, the user is distracted from his primary task. This can be very complicated for any location-based services, as any change of location usually needs audiovisual awareness and does not leave much room for interacting with a handheld device. The rating of importance of events is quite an open and interdisciplinary research topic involving computer science as well as social sciences and human–computer interaction research.

7.4 Further Reading

Event-driven architectures have seen a long tradition in computer science for their flexibility and for that they naturally fit a lot of application domains in which most interactions are isolated in time and can therefore be naturally mapped to events.

A nice overview about event-driven architectures has been given by Michelson in [2]. With respect to activity recognition which provides a lot of nontrivial events for indoor location-based services, the overview article of Turaga is recommended [4]. Another example of explicitly using events to model the behavior of a location-based service is given by Naguib and Coulouris in [3].

References

1. Downs, R.M., Stea, D.: Kognitive Karten. Harper & Row (1982)
2. Michelson, B.M.: Event-Driven Architecture Overview, vol. 2. Patricia Seybold Group (2006)
3. Naguib, H., Coulouris, G.: Location information management. In: Ubicomp 2001: Ubiquitous Computing, pp. 35–41. Springer, Berlin (2001)
4. Turaga, P., Chellappa, R., Subrahmanian, V.S., Udrea, O.: Machine recognition of human activities: a survey. IEEE Trans. Circuits Syst. Video Technol. **18**(11), 1473–1488 (2008)

Chapter 8
Simultaneous Localization and Mapping in Buildings

The only way to figure out whether our predictions have skill is to track them in real time, not retrospectively, against a naive baseline.

Roger Pielke, Jr.

Simultaneous localization and mapping stems from robotics and is a name for algorithms with which an autonomous robot can construct a map of its surroundings and derive its current position inside this map. Hence, the vast number of possibilities with which a robot can detect the physical world induces a vast number of algorithms with which a SLAM system can be constructed. But still, all SLAM algorithms have one central element in common: they must be able to solve the *data association problem*. The observation of the physical world is usually local. Only features of the physical world that are near to the robot can be experienced by the robot's sensory. Then, these observations form a time series of observations and the central ingredient into SLAM is always a mechanism to find out that at some point in time in the time series of measurements, the experience of the robot is the same as in another time. In these cases, the spatial trajectory forms a loop and the accumulated errors inside the sensor stream can be removed.

The typical high-level flow of a SLAM system is given in Fig. 8.1. First of all, the mobile device needs some sensory to experience the physical surroundings. These sensors can include bumpers, laser scanners, depth cameras, video cameras, stereo cameras, and the like. From these sensors, two data streams need to be extracted: one stream containing features which are suitable to detect loops and another data stream which is sufficient for pose and location estimation. The solution to the association problem is often called *loop detection* as the spatial trajectory forms a loop when the association algorithm finds out that the sensory detects identical surroundings at a specific instant in time. For the upper part of the SLAM flow, which involves estimating a trajectory, any positioning system can be used, even those which suffer from accumulating errors. It is, however, very common to use some sort of inertial navigation or odometry to derive the conjectured trajectory.

The top row of Fig. 8.2 depicts five time instants and values of a possible feature stream at these instants in time. It is easily seen that these features are highly different except at the first and the fifth time instants. The loop detection algorithm will find this correspondence of one and five. Below this data stream, the conjectured

© Springer International Publishing Switzerland 2014
M. Werner, *Indoor Location-Based Services*, DOI 10.1007/978-3-319-10699-1_8

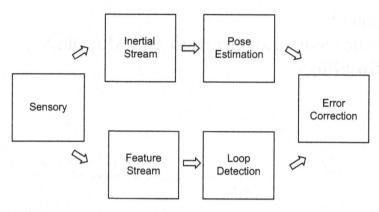

Fig. 8.1 General outline of SLAM algorithms

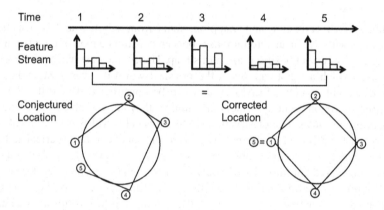

Fig. 8.2 Principle of loop detection

trajectory is depicted, which is generated on the fly from all sensor measurements. When a strong correspondence in the feature stream is detected, then the SLAM framework assumes that a spatial loop has been closed. This means that the locations at time instants one and five coincide. In the step of error correction, the actual error of the conjectured trajectory at time instant five is known and can, for example, be scattered along the points one to five assuming that each contributed the same amount to the final error vector.

8.1 Data Sources for SLAM

The selection and combination of algorithms in a SLAM framework depend highly on the type of data that is available and on the type of features that are extracted from it. However, it should be clear that the data must suffice to estimate the location (or pose) at any time with allowed, but limited, errors. Furthermore, it should enable some algorithm to find correspondences and an error propagation model to correct the conjectured trajectory given such correspondences. Before some important algorithms are presented in Sect. 8.2, we shortly review the most important types of data streams that are being used in SLAM applications.

In general, SLAM applications perform two things in parallel: positioning and mapping. The following sections present four common data sources that often appear in SLAM algorithms. The first two have been selected as they are among the most common techniques to directly measure the movement and directly measure the environmental features in robotics applications. The other two data types have been selected due to their availability in smartphones and consumer electronics.

8.1.1   Data from Inertial Navigation Systems

One of the important aspects of SLAM is the fact that the system has to calculate an estimate of the current location and orientation in space without interruption. This is often achieved by using an inertial navigation system (INS). INSs rely on estimating the movement relative to a known previous location and hence suffer from error accumulation. As presented in Sect. 3.3.5, this type of position update vectors can be constructed from summing up measurements of acceleration, estimating orientation, counting steps, or steering information in the case of autonomous robots and vehicles. INSs usually rely on summing up these estimated and erroneous position update vectors and hence have accumulating errors. However, in a SLAM application these errors can be overcome during the loop detection and error correction steps. For the purpose of using inertial navigation in SLAM, the INS should provide a single estimate of the location at any point in time based on the previous known state of the system.

8.1.2   Data from Laser Scanners

Laser scanning and LiDAR provide pretty accurate estimates of the distances between a mobile device and the environment from a given unknown point of view. For 2D-LiDAR, the system usually provides a vector containing the distance to some object inside the environment in one specific direction relative to the viewing direction.

Figure 8.3 depicts a scene including a robot and a common 2D-LiDAR laser scanning result. This two-dimensional approach is, of course, only sensible when the device orientation is fixed in at least one rotational axis of the three-dimensional world. Hence, it applies best to vehicles and robots limited to two-dimensional movement.

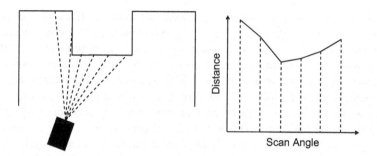

Fig. 8.3 2D laser scanning inside buildings

8.1.3 Data from Landmarks

A landmark is defined to be a feature of the environment, which is distinctive at least relative to a given size of vicinity. From this definition, it is clear that landmarks are perfectly suited for solving the association problem. However, it is not easy to extract natural landmarks from sensor streams. For dedicated applications and in research, however, people have brought out dedicated landmarks to the environment for solving the association problem in a simple and direct way. These landmarks can be visual codes such as barcodes or QR codes, active infrastructure elements such as Wi-Fi access points or Bluetooth beacons, NFC or RFID tags, or any other positioning infrastructure providing proximity services with sufficient accuracy. On the contrary, landmarks can also be extracted from the natural environment. Examples are visual object detection in images or audio fingerprints capturing characteristics of the background noise at different locations. In this case, of course, errors are introduced by mistakenly selecting non-unique sensor features as landmarks leading to wrong associations and, thereby, wrong assumptions of spatial loops. This type of error is called false positive: the system assumes an association though there is none. Another type of errors called false negatives also limits the performance of the SLAM system as a spatial loop closure is missed, and hence, error propagation can only be applied later with a greater inconsistency due to accumulation of errors.

8.1.4 Data from Camera Systems

It is clear that image recognition and matching can be used to detect predefined or automatically extracted landmarks from video cameras. However, cameras can also be used in a direct way to estimate the movement of the mobile device and hence provide position updates to the systems. In this area, the optical flow is among the most prevalent techniques. The optical flow of two consecutive images in a video stream is the vector field in which each pixel serves as the base of a vector pointing to the pixel in the second image, where the physical object represented by the pixel has moved.

The extraction of optical flow is usually not possible for each pixel, but only for those that have a sufficiently strong uniqueness to be correctly found again in the second image. Finding points of sufficient uniqueness in images or in pairs of images is the essential problem of feature point-based methods of image recognition.

8.2   Important Algorithms for SLAM Systems

This section includes a set of different algorithms for the various tasks which have to be accomplished by a SLAM system. We limit the exposition to the most widely adopted approaches. For each subproblem of SLAM, a lot of other algorithms exist.

As a first algorithm, we explain visual feature point extractions. This is a technique with which the similarity of objects located in camera-captured scenes can be calculated, and it can be applied to solve the pose estimation problem or the loop detection problem given a sequence of camera images. In order to solve the pose estimation problem from a camera stream, simpler algorithms exist, which detect the optical flow. The optical flow is the two-dimensional vector field of movements of points inside the image between consecutive frames. For building a map from these highly erroneous feature streams, the point cloud registration problem is often introduced. For several times, a point cloud is constructed representing the current and local view of the surroundings. The task of point cloud registration algorithms is given by finding the most probable transformation translating the first point cloud into the second point cloud. For a camera SLAM system, this defines the camera egomotion and finally solves the pose estimation problem. For point registration, the Iterative Closest Points (ICPs) approach is discussed, in which an initial guess is iteratively refined by locally optimizing discrepancies between one point cloud and the other point cloud transformed by the currently conjectured transformation. The Random Sample Consensus (RANSAC), instead, samples sufficient subsets from the available data in order to calculate a transformation and does this often. The best transformation based on a global error measure and calculated from a random sample is then used as the transformation. This approach is based on assuming that measurements are split into inliers and outliers and that the probability of once calculating a transformation from inliers is high enough. These two algorithms construct sequences of transformations including inevitable errors. Therefore, graph-based optimization algorithms are introduced, which can correct these errors based on a graph representation of the different poses and transformations.

8.2.1   Visual Feature Point Extraction

Visual feature point extraction is a camera vision technique in which algorithms select points depending on some local measure of interestingness and describe them with descriptors built from several numbers such that the Euclidean distance between these descriptors is able to capture similarities of the surroundings. Two widely adopted techniques of feature point-based image matching are given by the Scale Invariant Feature Transform (SIFT) and Speeded-Up Robust Features (SURF). In both cases, the definition of interesting points is given by local extrema in the gray scale image. In order to be able to capture different scales, the input images are scaled into several sizes called octaves, and matching is performed inside these octaves of images. An interesting point is some local extremum in the gray scale images, which is detected in several neighboring octaves. In a first descriptive step, an orientation and scale are assigned to the interest point depending on the local image gradients. Then the orientation and scale are used to fit a square grid over the image centered at the interesting point. For a grid of 64 cells, a 64-dimensional descriptor is then extracted by calculating the orientation inside each grid cell.

It has been shown that the Euclidean distance between such descriptors can well be used to detect equivalent points or objects inside different scenes. However, a noise rejection has to be applied to remove matches between points in different images, which are not distinctive. Therefore, for a feature point from the first image, the nearest and second-nearest feature points in the second image are calculated and the distance ratio has to exceed a predefined threshold. In the case of nondistinctive points, for example, generated from image noise, the ratio of the nearest and second-nearest match will be small, as a lot of nearly matching points exist in the second image. If it is high, it is considered a real match and finally the number of matches in one direction or in both directions indicates similarity of images.

Notably, the scale space approach together with the rotation invariance provided by estimating a global orientation of each feature point and describing the surroundings again by orientations leads to a completely scale- and rotation-invariant system. Therefore, limited differences in perspective or orientation of objects inside the scene can be overcome. This makes methods based on feature points more suitable for dynamic scenes as compared to edge detection techniques or others.

For SLAM algorithms, these feature points can be used to either solve the association problem by using, for example, the number of matches between two images as a score for their similarity or using a threshold to detect when these two images contain the same objects. Including the scale information of the descriptors or some spatial relationship between the pixel locations, the system can also try to guess parts of the camera pose. The feature points can also be used to detect the relative movement of feature points in consecutive images and calculate some measures such as the optical flow. However, simpler algorithms exist for this task as explained in the next section.

8.2.2 Optical Flow Estimation

For the problem of estimating the egomotion of the camera in the scene, the optical flow can often be used together with assumptions limiting the degrees of freedom of the camera. For example, when a camera is mounted to a vehicle, the optical flow can be used to estimate the speed. Recall that the optical flow of two consecutive images is a vector field connecting some pixel from the first image to their location in the second image. There are very different approaches to extracting this vector field based on computational considerations and the actual matching scheme. In most cases, the estimation of the optical flow relies on the assumption of constant intensity: two pixels stemming from the same object in the real world will have the same intensity. Usually, the pixels in two-dimensional optical flow estimation are represented by three coordinates: two spatial ones x and y and a time t. Constancy of the pixel intensity I can be expressed using symbols $\Delta x, \Delta y, \Delta t$ for the changes in the parameters as follows:

$$I(x, y, t) = I(x + \Delta x, y + \Delta y, t + \Delta t).$$

In general, the optical flow estimation relies on smallness of movement. It is customary to assume that the pixels in consecutive images have not moved too far. Due to the smallness of the offsets, we can apply a Taylor series expansion of order one using the three partial derivatives as follows:

$$I(x + \Delta x, y + \Delta y, t + \Delta t) \approx I(x, y, t) + \frac{\partial I}{\partial x} \Delta x + \frac{\partial I}{\partial y} \Delta y + \frac{\partial I}{\partial t} \Delta t + R.$$

Ignoring higher-order terms R, we can then require the sum of three partial derivatives to vanish:

$$\frac{\partial I}{\partial x} \Delta x + \frac{\partial I}{\partial y} \Delta y + \frac{\partial I}{\partial t} \Delta t = 0.$$

In this equation, the values of Δx and Δy are of interest, as they constitute the optical flow in this situation. To simplify this further, it is a good idea to divide the entire equation by Δt resulting in an expression about the velocities $\Delta x / \Delta t$ and $\Delta y / \Delta t$, hence a linear in two unknowns u and v representing pixel velocity vectors:

$$I_x \; I_y \genfrac{}{}{0pt}{}{u}{v} = I_t.$$

In this expression $I_x = \frac{\partial I}{\partial x}$ and similar for y and t. In general, this equation cannot be solved without additional constraints, and a vast number of different algorithms use different constraints to find a solution to this equation often referred to as the aperture problem of optical flow estimation.

This chapter will focus on the method of Lucas and Kanade [5]. This method introduces a third constraint to overcome the underdetermination of the system of equations above. This third constraint assumes that the surroundings of a moving pixel move together with this pixel. Therefore, a square window of window size l containing l^2 pixels around the pixel coordinate is being used for matching. Using the same reasoning as above assuming constancy of intensity and smallness of movement, we directly obtain an overdetermined system of linear equations. Let p_i be the points of the matching window. Then the system of equations can be given as

$$\underbrace{\begin{bmatrix} I_x(p_1) & I_y(p_1) \\ I_x(p_2) & I_y(p_2) \\ \vdots & \vdots \\ I_x(p_{25}) & I_y(p_{25}) \end{bmatrix}}_{A} \underbrace{\begin{bmatrix} u \\ v \end{bmatrix}}_{p} = - \underbrace{\begin{bmatrix} I_t(p_1) \\ I_t(p_2) \\ \vdots \\ I_t(p_{25}) \end{bmatrix}}_{b}.$$

This system of equations $Ap = b$ can be solved by using least squares regression introduced in Sect. 3.1.1 by solving the associated normal equation:

$$A^t Ap = A^t b.$$

A straightforward extension to the basic algorithm is given by using weighted least square regression in the region matching step. In this case, a weighting matrix is introduced to give a higher importance to pixels that are near to the center pixel and less importance to those pixels of the matching window, which are farther away. This leads to the following weighted normal equations using a square diagonal matrix W for weighting:

$$A^t WAp = A^t Wb.$$

Obviously, the whole approach will fail for sequences in which the matching region changes too much between consecutive frames or when the movement between consecutive frames is too large to use the Taylor series expansion. To reduce these effects, a scale space approach similar to the one used for visual feature detection in SIFT and SURF can be applied. In these cases, the matching is performed from coarse to fine where each level of the scale space takes the movement estimation of the previous level for initialization. This is the basis for the well-known Kanade–Lucas–Tomasi (KLT) feature matching algorithm. In summary, optical flow determination is usually based on the following two basic assumptions:

- The pixel intensity of a pixel in the first image and its correctly associated pixel in the second image does not change
- The pixel movement between two consecutive frames is small enough for the error of an order one Taylor series expansion to be neglected.

This leads to a linear equation with two unknowns, but only a single equation, and, hence, is not solvable without additional constraints. The method presented in the previous paragraph adds the assumption that the direct surrounding region of a pixel moves together with the pixel and translates this into a system of equations leading to a solvable problem.

A fundamentally different way to complete an algorithm for optical flow estimation is given by adding a global constraint about the two consecutive images. The well-known method of Horn and Schunck adds a global constraint of smoothness to the equation system. In this case, a measure of smoothness is given as a global functional on the image which is then minimized. The main advantage of such a global approach is that the flow is dense in the sense that also the inner of homogenous regions in frames contains flow vectors. In other words, every pixel of the first image is assigned a corresponding pixel location in the second image. On the other side, this method is quite sensitive to noise. Moreover, this sensitivity has a global effect and has a negative impact on all flow vectors. On the contrary, local methods can deal with noisy regions in most cases by just not providing a flow vector there, while regions with small amounts of noise lead to good flow vectors there.

8.2.3 Iterative Closest Points

ICP algorithms are among the most important methods to solve the registration problem of point clouds. The input to the ICP algorithm consists of two sets of points S and T, often taken from Euclidean vector spaces $S, T \subset R^d$ of equal dimension d. Their task is to find the optimal transformation ϕ from a set of rigid transformations. ICP finds applications in the area of 3D scanning, where the scanner returns a depth of field vector or matrix. Then ICP can be used to register different views into a consistent coordinate system resulting in a global point cloud containing all points from different laser scans. These can then be processed to generate, for example, three-dimensional models of the surroundings. In summary, ICP algorithms are suited to solve the following registration problem of point clouds: *find the coordinate transformation that transforms two point clouds into each other with minimum error.* Basically, ICP starts with an initial guess of the transformation transforming a point set into another. It then uses this guess to find reasonable matches between points and modify the transformation to better fit these matches. The initial guess can be derived from a variety of external information sources. However, in SLAM applications it is obviously possible to use the position estimation part for this purpose, when ICP shall be used for loop detection. A lot of ICP algorithms can be classified according to the following six steps and their various implementation details [9]. Figure 8.4 depicts these steps in visual form.

It is often computationally infeasible to calculate a matching for all available points. When matching geometry descriptions containing solid forms like spheres, for example, the number of available points for matching is even infinite. Hence,

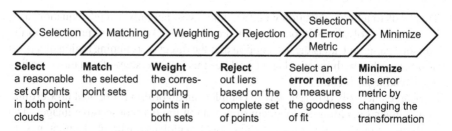

Select	**Match**	**Weight**	**Reject**	Select an	**Minimize**
a reasonable	the selected	the corres-	out liers	**error metric**	this error
set of points	point sets	ponding	based on the	to measure	metric by
in both point-		points in	complete set	the goodness	changing the
clouds		both sets	of points	of fit	transformation

Fig. 8.4 General flow of ICP algorithms

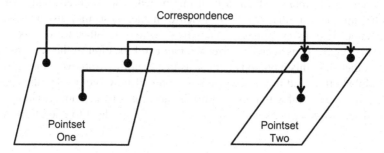

Fig. 8.5 Point matching example for ICP algorithms

from the available points given either by measurement or by sampling of geometrical directives, a suitably sized subset needs to be extracted in the *selection* step
of ICP algorithms. These point sets should be large enough to actually capture
enough information about the actual transformation between the datasets including
some overdetermination to counterfeit noise in cases where the data comes from
measurements. On the opposite site, the smaller the set is, the faster the remaining
steps can be achieved. In the second step, the selected points are matched. Therefore,
each point in the first point cloud is assigned to a point in the second point cloud.
This assignment is depicted in Fig. 8.5.

For this step, the initial guess of transformation is usually used and some sort
of selection of nearest points is performed giving reason to the term *closest* in the
name ICP algorithm. The third step of *weighting* can be used to assign importance or
confidence estimates to the individual point correspondences. These values can be
used for the fourth step of *rejecting* obviously wrong assignments before calculating
further. In the fifth step, a suitable error metric for the given registration problem
can be chosen. Based on the characteristics of errors, different choices might be
sensible and can lead to different results for the same data. The last step consists of
minimizing the chosen error metric. Therefore, direct methods can be used, which
find the optimal value for the current data in a single step. Often, however, only
coarse corrections such as gradient descent are being used to update the initial
transformation guess. With this updated transformation, the process gets iterated in
whole or in parts until a sufficient breaking condition or some computational time
limit is reached.

8.2.3.1   An ICP Example

In a paper of Rusinkiewicz, different algorithms for the subtasks of ICP are evaluated against the following baseline approach which will be explained in detail in this section [9]. It has also been proposed by other authors and is often only slightly different from the numerous variants of ICP in the literature. It is not intended to represent the best practice or most up-to-date research result. Instead, it shall serve as a detailed example from which application-optimized ICP variants could be derived easily. For the selection of points, a random sampling of the source point cloud is used. This limits the computational complexity of the following steps while keeping a representative subset of the points. When iterating this process, another subset of points can be sampled such that the number of source points used in the calculation increases over time. The second step of matching points is usually performed in a next neighbor manner. Therefore, each point of the source set is transformed by the current estimate of transformation and assigned to the nearest point in the target point set. This, however, introduces a lot of errors from different sources. As a result, a number of techniques have been proposed to reduce the number of wrong matches which could mislead the remaining steps of ICP. To reduce these wrong matches, additional features of the points can be compared in a thresholding manner including color or normal vectors. These points are then called compatible points, and the initial assignment is filtered to include only compatible points. Without additional information, matches with heavily increased distance in the assignments are often rejected as they might stem from occluded pixel information or wrong measurements. Furthermore, neighborhood relations can be used to constraint assignments. Therefore, assignments for which the distance to neighbors is increasing too much are rejected. It should be noted that finding the k nearest neighbors is a complex problem. The naive approach for a single point p from the source point cloud S consists of calculating the distance between all points of the second point set $T = \{t_i\}_{i=1...N}$ and the transformation ϕp of the source point remembering the nearest point. For point clouds in Euclidean spaces, the following expression gives the algorithm of linear search, which consists of Nd operations, where d is the number of dimensions of the target vector space $T \subset \mathbb{R}^d$

$$\text{NN}(\phi p, T) = \text{argmin}_{i=1...N} \|\phi p - t_i\|.$$

It is possible to increase performance by using spatial indices such as the kD-tree as explained in Sect. 4.5. However, it is not always clear whether the time used for constructing these spatial indices pays off in a specific situation. Another approach to speed up nearest neighbor matching is given by locality-sensitive hashing in which a constant complexity function of the points can be used to assign numbers to each point such that the probability is high that points with the same number have low distance. This technique is explained in Sect. 6.3.3 of this book in the context of trajectory classification. For the current example, we do not use any weighting. However, it is clear that, for example, color distances or other features can easily be incorporated into this step unless they have been used for distance calculation in the nearest neighbor search. For rejection of wrong matches, we use a combination of the following two heuristics: first of all, we remove a fixed quantile of largest distance matches. Hence, for all assignments, we calculate the assignment error vector length and remove the most distant matchings. For an error metric, we select the sum of the squared distances of the projected points and the assigned points. Let p_i denote the point in the second point cloud, which has been assigned to t_i, then we use the error measure E as follows:

$$E = \sum_i \|\phi p_i - t_i\|^2.$$

This is a reasonable choice, as the last step can then be expressed in closed form: there exist closed form solutions for determining the optimal transformation ϕ^* with respect to this error metric given the assignments from the current transformation ϕ. For each round, the complete process including random source point selection is repeated. This completes an iterative algorithm finding better and better transformations aligning the point clouds with each other. Such iterative approaches have the nice feature that they can be either stopped when converging or after a predefined time. Hence, they are suited for real-time systems in which the computation time for each frame alignment is limited.

8.2.4 Random Sample Consensus

RANSAC is another computer algorithm that can be used for point cloud registration. Though it has a high-level flow very similar to ICP, it is based on a completely different philosophy. It is based on the assumption that the dataset consists of two types of data elements: inliers and outliers. In general, an outlier is an observation, which significantly differs from other observations. These outliers are usually generated from measurement errors. Inliers are the opposite: every element that is not an outlier is an inlier. RANSAC algorithms are based on the assumption that the outlier probability P_o for a dataset is known. The overall flow of a RANSAC application is as depicted in the flow diagram in Fig. 8.6.

For as many iterations as dictated by some criterion, select a small subset of the data large enough to build a model from it under the assumption that it does

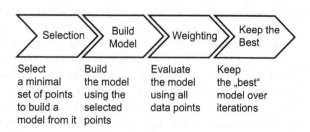

Fig. 8.6 General outline of RANSAC algorithms

Selection	Build Model	Weighting	Keep the Best
Select a minimal set of points to build a model from it	Build the model using the selected points	Evaluate the model using all data points	Keep the „best" model over iterations

not contain outliers. In the second step, calculate a global error function for this model incorporating all data elements. If this error function is small enough, add all elements that can be explained by the model to a set of inlier datapoints for this model. It is customary to define explanation by a simple threshold: all data elements that are nearer to the prediction of the model than a threshold value τ are inliers. This set is usually called consensus set. This process is iterated and the best model together with its consensus set is remembered. When this iteration ends, one can construct a new model from the set of all inlier elements for the best model found so far. This can usually be achieved by using least squares or other optimization technique to overcome inconsistencies in the data due to measurement noise. A central parameter of this algorithm is, of course, the number of iterations n to be performed. This number can be calculated from knowing a good estimate of the fraction of outliers f_o in the dataset and an intended probability of success p. This probability of success p is the probability that one of the n iterations was able to find a set consisting only of inliers. Identifying probabilities and fractions for ease of exposition, the probability of choosing a single inlier is given by $1 - f_o$ and the probability of independently selecting a subset of size k consisting entirely of inliers is given by

$$p_i = (1 - f_o)^k.$$

As p_i is the probability of selecting a set consisting only of inliers, $1 - p_i$ is the probability that at least one outlier is in the set. The probability that the algorithm never selects a set consisting only of inliers is then given by the power

$$(1 - p_i)^n.$$

The probability p of success is the opposite of the case that the algorithm never selected a full inlier set. Hence,

$$p = 1 - (1 - p_i)^n$$

which can easily be solved for n by moving the leading 1 to the left side and applying the logarithm to both sides resulting in

$$n = \frac{\log(1 - p)}{\log(1 - p_i)}.$$

Hence, by defining an intended probability of success p and knowing p_i which is based on the fraction of inliers and the set size for the sampling step, one can calculate the number of iterations needed. Note that the number n is an upper limit in practical cases. This is due to the fact that we have randomly selected elements though it might be more common to select a set of pairwise distinct elements in the sampling step. This decreases the probability p_i of choosing a set of k inliers by decreasing the absolute value of the denominator and, hence, increases the number

Fig. 8.7 Number of
iterations needed in RANSAC
algorithms for a fixed success
probability and varying
fraction of outlier points

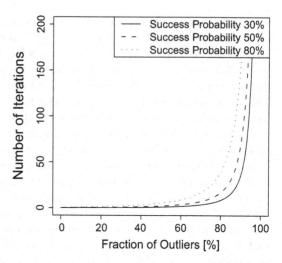

of needed iterations. Note that the arguments to the logarithm are between 0 and 1 and, hence, the logarithm is a negative number. The sign cancels, and hence, n increases as to be expected.

For a simple application example of matching a single line, one needs two points to find a model of a line, that is, $k = 2$. For this case, Fig. 8.7 provides an overview of iterations given an outlier fraction on the X-axis and different success probabilities for the different lines. The Y-axis denotes the number of iterations. These figures make clear that the number of iterations depends exponentially on the quality of the dataset.

But one can also see the stability with respect to outlier fraction. Even when 80% of the data are outliers, the system works with less than 150 iterations. As the model building step for a line from two points is rather trivial, this results in a quite performant algorithm. In general, the success of RANSAC algorithms is guided by the assumption that the fraction of outliers is known and sufficiently small to select inlier subsets from the dataset. There is a central trade-off to control the performance of RANSAC: increasing the size of the sampled sets might generate better models from data in the presence of noise. Unfortunately, also the probability of inliers in this dataset is increased.

8.2.5   Graph-Based Optimization Algorithms

A SLAM system is able to track the trajectory of the moving object any time, but possibly with errors. Moreover, a SLAM system has some possibility to generate associations between observations. These algorithms are often called the SLAM front end as they are usually based on direct observations of the surroundings and sensor data. This part can also be seen as the localization part of SLAM in two aspects: firstly, a coarse location is estimated from the sensor stream, e.g., by inertial navigation, and secondly, the data association problem can be seen as a proximity localization: find cases from the sensor stream where a similar location and pose of the mobile device should be assumed. From these algorithmic parts of SLAM, a single problem can be formulated in graph-based SLAM approaches, which aims to provide a consistent map. This is usually called SLAM backend. A SLAM backend shall find the configuration of poses and transformation, which maximizes the likelihood of the observations. This part can be seen as the mapping part of SLAM in which the observations are combined into a consistent representation. For this graph-based SLAM backend, the data of the front end is modeled as a graph together with a set of constraints. The graph is constructed from a stream of location or pose estimations of the mobile entity as provided by the SLAM front end. The vertices of this graph are annotated with the estimated pose (location, orientation, viewing direction, etc.) at a specific instant in time, and the edges between two vertices are annotated with the transformation information transforming the one pose into the other pose. As a third piece of information, a set of constraints is constructed from the results of the association problem: a constraint is a linking between two vertices in the graph, which represent the same (or similar) location. The task of graph optimization algorithms is to find the most probable graph, which best explains the measurements and fulfills the constraints. Figure 8.8 depicts the situation of a simple loop for a SLAM algorithm.

The true trajectory is drawn with a solid line and is not known to the system. The INS drifts a bit towards the left from the true trajectory. It is drawn with a dashed line and the circles represent the sample locations. These provide the vertices of the graph. The data association algorithms provided a constraint drawn as a dotted line between the second vertex of this trajectory and the last one in the figure. For this case, the graph optimization algorithm or SLAM backend will modify the trajectory such that these two distant vertices become near to each other while the trajectory still explains the measurements.

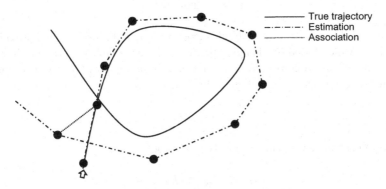

Fig. 8.8 Example of the graph-based SLAM optimization problem

8.2.5.1   Toro Algorithm

The tree-based network optimizer (TORO) algorithm is a widely used approach to solve the graph optimization problem of simultaneous localization and mapping systems. It belongs to the class of maximum likelihood (ML) methods as it is based on maximizing the likelihood of the observations. Let x denote the vectors assigned to each vertex in the graph. In many cases, this vector will contain three spatial coordinates as well as three angles completely defining the pose of the mobile entity. Let further δ_{ji} and Ω_{ji} denote the mean and the information matrix of an observation of node j as seen from node i. Let further f_{ji} compute a noiseless observation according to the current location of the vertices in the graph. Then the discrepancy between the observation δ_{ji} and the current map f_{ji} is given by the residual r_{ji} defined by

$$r_{ji}(x) = -e_{ji}(x) = -\left(f_{ji}(x) - \delta_{ji}\right).$$

If the observation and the current graph model were completely correct, the residuum would be zero. However, due to noise and possible wrong associations, this is impossible in reality. Therefore, one is interested in minimizing the residuum somehow. Therefore, in TORO, the impact of wrong assignments is calculated as the negative log likelihood of the observations incorporating the information matrix Ω_{ji} and assuming normal distribution of errors as well as independence of different constraints. Given a set $\mathscr{C}$ of constraints between vertices $\langle j, i \rangle$, the objective function is given by

$$F(x) = \sum_{\langle j,i \rangle \in \mathscr{C}} r_{ji}(x)^t \Omega_{ji} r_{ji}(x).$$

This error function is then minimized by computing the result x^* of

$$x^* = \mathrm{argmin}_x F(x).$$

For this problem, a lot of possibilities exist. Olson et al., for example, provide an iterative method based on changing the network constraint by constraint. In every step, they choose a constraint $\langle j, i \rangle$ and modify the graph in a way that aims to minimize its impact on the sum given by

$$r_{ji}(x)^t \Omega_{ji} r_{ji}(x).$$

Therefore, an update equation is easily written out

$$x_{k+1} = x_k + \lambda K J_{ji}^T \Omega_{ji} r_{ji}.$$

In this equation, the most important things are two: first of all, the direction of correction is essentially given by the residual. Of course, the error is reduced when moving into that direction. However, it is not clear how far to go. Hence, the rest of the formula gives an estimate about how far to go based on the local geometry (represented by K and the Jacobian matrix J) of the error surface. In this case, this estimate of distance is based on several things: the relaxation coefficient λ controls convergence as it is decreased for every iteration of this approach. In general, a larger value of λ amounts to a wider step in this algorithm. The matrix J_{ji} is the Jacobian matrix of f_{ji} and K is a preconditioning matrix. More information on the details of this geometric scaling approach can be found in the original paper on TORO [4] or on Olson's algorithm [7]. For our purpose, it should suffice to get a feeling that TORO reduces the global error function F summand by summand and step by step considering local geometry of errors. Of course, this algorithm is local in nature. In every step, only one single constraint is used to update the network. Therefore, each step can impair another step already taken. This is another reason why a relaxation parameter is needed. It limits the amount of change in the network after a fixed number of iterations. Therefore, the network will result near to an equilibrium state after several iterations which provides the final result of the TORO algorithm.

8.2.5.2 Hog-Man

The Hog-Man algorithm is another algorithm quite similar to the TORO algorithm. It is also based on minimizing the error function as given by

$$F(x) = \sum_{<j,i> \in \mathscr{C}} r_{ji}(x)^t \, \Omega_{ji} r_{ji}(x).$$

In total, the Hog-Man algorithm is based on Gauss–Newton method, which has already been used for localization in lateration and angulation in Sect. 3.1. Therefore, using an initial pose estimate, the error function is expanded to the first-order Taylor expansion (called local linearization) around this estimation, and finally, an update vector is calculated by solving a linear system. As the matrix in this linear system will be sparse by construction (it only has nonzero entries, where graph vertices are connected by constraints), a very efficient factorization can be used to solve this linear equation system. This factorization is called sparse Cholesky factorization. To that point, there is no central new idea behind the Hog-Man algorithm. We have an objective function and optimize it by differentiation of the error plugged into that equation. What is now the great inventive step of Hog-Man is a useful treatment of geometry. The pose usually consists of entries in a Euclidean vector space (e.g., the spatial coordinates) as well as elements from the different rotational groups $SO(2)$ or $SO(3)$. The fact that these groups are not Euclidean can lead to bad linearization in the classical approach due to singularities. In Hog-Man a manifold structure is used to overcome singularities. A manifold is a geometric object in which every point has a limited surrounding which is Euclidean. The Hog-Man algorithm is now based on defining adding operations, which enable the algorithm to avoid problems with singularities in the angles. In general, small update vectors are calculated in a Euclidean neighborhood of the initial estimate using the manifold structure, while the aggregation of these small individual offsets can pass by singularities with less numerical problems.

8.3 Several Well-Known SLAM Approaches

This section will briefly introduce some real SLAM systems. The early systems all stem from the robotics domain and are often used to build an occupancy grid map of the surroundings of a moving vehicle equipped with a LiDAR-based laser scanner obtaining a depth measurement from the actual pose of the mobile robot into its current viewing direction as depicted in Fig. 8.3. These systems are the basis for SLAM for indoor location-based services. However, the arbitrary motion and the large errors in low-cost IMUs of smartphones actually do not yet suffice for SLAM approaches directly. Therefore, the SLAM approaches are often integrated with error filtering methods as the following examples show.

8.3.1 Extended-Kalman-Filter SLAM

Extended-Kalman-Filter SLAM was one of the most common SLAM algorithms until the invention of FastSLAM. It is based on the extended Kalman filter (EKF) (Sect. 5.4), which assumes Gaussian system noise and linearizes nonlinear system models by Taylor approximation such as to suit the classical Kalman filter algorithm. The state at time instant k of EKF-SLAM is usually expressed as a single global state vector S containing the pose of the robot as well as the locations of the observed landmarks. For a two-dimensional case, this amounts to a vector

$$S(k) = (x(k), y(k), \phi_k, x_1, y_1, x_2, \ldots, x_N, y_N)^t.$$

Note that only the pose of the mobile entity has a parameter indicating time. This is due to the fact that landmarks are usually modeled stationary in EKF-SLAM. The motion of the mobile entity is then modeled by an equation incorporating the values $x(k)$, $y(k)$, and $\phi(k)$. Assume that the time between consecutive time instants is given by Δt. The following set of equations describes a vehicle with a steering angle $\gamma(k)$ and velocity $v(k)$. These parameters are assumed to stay constant during a timestep of Δ_t. The value B describes the wheelbase, that is, the distance between the front axis and the rear axis

$$x(k+1) = x(k) + v(k)\cos(\phi(k) + \gamma(k))\Delta t$$

$$y(k+1) = y(k) + v(k)\sin(\phi(k) + \gamma(k))\Delta t$$

$$\phi(k+1) = \phi(k) + \frac{v(k)\Delta t}{B}\sin(\gamma(k)).$$

Assuming that the mobile vehicle is able to measure the distance and relative angle to a specific landmark, the observation model can be given by calculating the current distance from the state to the landmark and the expected angle calculated based on the current state:

$$z_i(k) = (d_i(k), \psi_i(k)) = \left(\sqrt{(x_i - x(k))^2 + (y_i - y(k))^2}, \arctan \frac{y_i - y(k)}{x_i - x(k)} - \phi(k) \right).$$

If this observed landmark does not fit to a landmark already in the map, a new landmark is added and the state vector grows. The most important problem of EKF SLAM is induced by the assumption of Gaussian noise and that it is quite sensitive to heading variance [2].

8.3.2 FastSLAM

FastSLAM is based on EKF-SLAM and addresses a scalability issue of EKF SLAM. For EKF implementations of SLAM, the update needs time quadratic in the number N of landmarks. This is due to the fact that the Kalman filter maintains covariance matrices of size $O(N^2)$, which can all change due to a single landmark observation. This effectively limits the number of landmarks to a few hundreds depending on the actual computational power available, while for many SLAM applications millions of features are observed. The key idea of FastSLAM is the observation that—assuming that the true track is known—the landmark estimation problem is completely independent: every landmark can be located independent of the location of the other landmarks. Basically, the FastSLAM approach maintains a set of particles each representing a trajectory conjecture and for each particle, a set of N Kalman filters keeping track of the location estimate of the landmarks. Using a tree-based data structure, the overall complexity can be reduced to $O(M \log N)$ where M denotes the number of particles in the filter system [6]. The central result of FastSLAM is its better runtime. This runtime allows SLAM systems to track an order of magnitude more landmarks and, hence, provide better maps in shorter time.

8.3.3 Grid-SLAM

Grid-SLAM is a combination of FastSLAM for laser-based distance measurements and scan matching. It integrates FastSLAM with an occupancy grid map representation. Furthermore, it integrates the sequence of laser range measurements into the movement estimation of the mobile entity. This corrected odometry information together with the remaining laser scan data is then used to build the map. The incorporation of laser ranging information into the movement estimation allows the SLAM system to close even larger loops which has been impossible with different approaches.

8.4 Summary

Simultaneous localization and mapping stems from the domain of self-orienting robots. The tight integration of localization and mapping leads to systems that, hopefully, converge to a maturity which enables crowdsourcing of environmental models without user attention. However, classical approaches to SLAM rely on relatively reliable sensory including laser scanners and high-precision INSs. Though some sort of inertial movement estimation is possible with a smartphone using accelerometers and gyroscopes, it is difficult to reach a level in which the inertial errors are small enough to be corrected by a loop detection system. Moreover, smartphones are typically unable to capture sensible video streams while people are moving around. Therefore, a technique for finding associations in sensor streams is still missing. However, a hybrid approach can be of great potential: using SLAM-based systems, possibly with special sensory or at least with people holding a camera into their viewing direction, an environmental model could be generated by volunteers, which might be used in order to provide location-based services inside buildings to the public. Especially the integration of mobile infrastructure components such as cleaning machines into an ecosystem of indoor location-based services can compensate for database drift problems and simplify the generation of environmental models.

8.5 Further Reading

A good overview of the techniques of simultaneous localization and mapping is given in a survey of Aulinas et al. [1]. Another overview article is given by Thrun and Leonard [10]. Furthermore, the article introducing RANSAC is worth reading [3]. Furthermore, I propose to have a look on up-to-date libraries and projects providing implementations and novel algorithms such as OpenSLAM.org which includes a large amount of information on different aspects of SLAM [8].

References

1. Aulinas, J., Petillot, Y.R., Salvi, J., Lladó, X.: The slam problem: a survey. In: CCIA, pp. 363–371. Citeseer (2008)
2. Bailey, T., Nieto, J., Guivant, J., Stevens, M., Nebot, E.: Consistency of the ekf-slam algorithm. In: 2006 IEEE/RSJ International Conference on Intelligent Robots and Systems, pp. 3562–3568. IEEE, New York (2006)
3. Fischler, M.A., Bolles, R.C.: Random sample consensus: a paradigm for model fitting with applications to image analysis and automated cartography. Commun. ACM 24(6), 381–395 (1981)
4. Grisetti, G., Stachniss, C., Burgard, W.: Nonlinear constraint network optimization for efficient map learning. IEEE Trans. Intell. Transp. Syst. 10(3), 428–439 (2009)

5. Lucas, B.D., Kanade, T., et al.: An iterative image registration technique with an application to stereo vision. In: IJCAI, vol. 81, pp. 674–679 (1981)
6. Montemerlo, M., Thrun, S., Koller, D., Wegbreit, B., et al.: Fastslam: a factored solution to the simultaneous localization and mapping problem. In: AAAI/IAAI, pp. 593–598 (2002)
7. Olson, E., Leonard, J., Teller, S.: Fast iterative alignment of pose graphs with poor initial estimates. In: Proceedings 2006 IEEE International Conference on Robotics and Automation, 2006. ICRA 2006, pp. 2262–2269. IEEE, New York (2006)
8. OpenSLAM: Give your algorithms to the community. Online (2014). http://www.openslam.org
9. Rusinkiewicz, S., Levoy, M.: Efficient variants of the icp algorithm. In: Third International Conference on 3-D Digital Imaging and Modeling, 2001. Proceedings, pp. 145–152. IEEE, New York (2001)
10. Thrun, S., Leonard, J.J.: Simultaneous localization and mapping. In: Springer Handbook of Robotics, pp. 871–889. Springer, Berlin/Heidelberg (2008)

Chapter 9
Privacy and Security Considerations

Privacy is effective ability to misrepresent yourself.

Dan Geer

Ubiquitous computing systems including location- and context-aware systems in general raise privacy concerns. Due to the systems becoming smart and being supposed to support the user in his actual situation, these systems thus collect information about user situations. A leakage of this information to other people or computer systems is often felt to be objectionable. The most important risk scenario in computer system is given due to the virtually unlimited ability to store and query such situation data in the future.

In order to understand the inherent privacy threats in situation-aware computing, the following example is very useful considering a classical trusted peer: A private secretary, for example, will collect private and compromising information during everyday work. However, a secretary is a human being and the loyalty of a secretary cannot be compromised by a security leak. Furthermore, the amount of work to corrupt the secretary is high and grows with the number of targeted individuals. If a computer system takes over the task of a secretary, e.g., planning trips and managing the calendar, the computer system will collect and use the same amount of private and compromising information for his job. However, it can easily leak due to security weaknesses and can easily be queried nearly independent of the amount of compromised individuals. In the classical example, of course, an attacker can also try to gain compromising information through the secretary. However, the involved strategies such as blackmail and bribery are expensive and risky for an attacker. Exploiting security weaknesses of computer systems is comparably easy.

This example also illustrates the tight interconnection between privacy and security, which often are mistaken or mingled. Privacy is best viewed from an information perspective: some information is declared as private, and a system provides privacy if this private information is kept inside a bounded domain. Security, however, is best seen as a property of a system that it does not allow an attacker to do anything unless explicitly authorized. These two concepts cannot exist without each other. A secure system architecture is a precondition for privacy protection. Unfortunately, privacy is often discussed with respect to an honest adversary, which does not include attacks to the computer system.

© Springer International Publishing Switzerland 2014
M. Werner, *Indoor Location-Based Services*, DOI 10.1007/978-3-319-10699-1__9

Privacy is in general term, which is difficult to grasp. Before the computer age, privacy was often a very personal relation: the amount of sensitive information given to another person was bound to the level of trust in this person. From this context, a disclosure of sensitive information was often due to a wrong assumption about the trustworthiness of a person. Still, these leaks are local in nature and can be controlled by law. In the computer age, however, privacy concerns come up due to the ability of computer systems to collect and store data without having the ability to warrant protection of these data collections against fraud and theft in practice. Computer systems face viruses and security flaws in both software and hardware.

Moreover, privacy protection does not include the protection of already-known facts. This introduces severe difficulties to define privacy with respect to nontrivial attacker models. The best suited notion is differential in nature and tries to limit the amount of information an attacker can add to his already existing knowledge by a query or system. As a simple example, think of the number of male people in the world. Most people will assume that the number of male is approximately the same as the number of female people in the world. Hence, when the amount of male people in some city is published, the amount of privacy loss for the citizen is given by the difference between the general assumption and the published information. Publishing data about a city where only male people live, however, results in a lot of information for an attacker as there are only very few places in the world where no women live.

In general, two architectures for privacy-protecting systems have been proposed for computer systems: the introduction of *trusted third parties* and the *distribution* of sensitive information over different entities.

Introducing a trusted third party is a very simple approach. In this case, all sensitive information is disclosed by the users to a single trusted party, which takes care of this data. This trusted party can perform calculations and provide interfaces to services relying on this sensitive information without exposing the linking of user and sensitive information to a service. A trusted third party is also a very popular approach as it can be used as a platform for service integration and for performing queries. However, trusted third parties do not introduce any privacy into the system unless they use specific approaches during brokerage of queries between users and services. These approaches are of great interest and can sometimes even be applied without a trusted third party.

A lot of people argue that the introduction of a trusted third party is an unrealistic simplification to reality and, hence, does not help in any way. It introduces a single point of attack, a large collection of sensitive information, and does not scale with the number of users. These arguments apply to simple instantiations of trusted third parties as servers and databases. Conceptually, however, a trusted third party can also be realized in a scalable or distributed manner. It should be seen as a concept rather than a single entity or party.

What is true about a trusted third platform is the problem of trustworthiness. How should a trusted third party become trustworthy? Firstly, the trusted platform becomes valuable due to the value of the data it processes. The service providers need a flexible and generic interface and the trusted platform will have to collect all data as it cannot assess the utility of a single piece of information with respect to external services the platform does not know. The utility for a service provider is linked with the accuracy and precision of the result of a privacy-preserving query with respect to the targeted individual.

Another concept for privacy-preserving systems is given by distribution and distributed computation. If the sensitive information does not exist in one common logical place, it cannot be leaked to an adversary in masses and the attack surface can be arbitrarily large rendering actual attacks infeasible. This is the case for the example of compromising the privacy of many people through their respective secretaries. Another very common example due to Andrew Yao is given by the millionaire's problem explained in the next section on multiparty computation.

Today, a lot of different definitions of privacy are used and consequently a lot of different algorithm providing privacy have been proposed. Inside this book, we will concentrate on the basic mechanisms, which are applicable to location-based services and do not follow a conceptual construction of the involved privacy definitions. This would easily fill textbooks centered on privacy.

9.1 Multiparty Computation

In 1982, Andrew Yao gave the following simple problem and raised the associated question:

> Two millionaires wish to know who is richer; however, they do not want to find out inadvertently any additional information about each other's wealth. How can they carry such conversation? [8]

This problem is a specific instantiation of a more general problem of evaluating an integer-valued function $f(x_1 \ldots x_N)$ in a distributed system in which each value x_i is known solely by node i, taken from some bounded integer domain, and should not be disclosed to any other node. The example above is then given by calculating the Boolean function $f(a, b) = (a > b)$ of two nodes.

One of the protocols originally proposed by Yaw is given by the following approach for two millionaires Alice and Bob:

Assume that Alice owns i million dollar and Bob owns j million dollar from the domain

$$1 < i, j < 10.$$

Assume that Alice and Bob each have a public key E_a and E_b, respectively. Let then Bob pick a random number x of N bits. Bob encrypts this value using Alice's public key, $k = \mathcal{E}_a(x)$. Bob then transmits the number $k - j + 1$, which contains his own amount of money indistinguishably from an effectively random number k. Using the bounded integer domain, Alice can compute privately the numbers

$$y_u = \mathcal{D}_a(k - j + i)$$

by adding $u \in 0 \ldots 9$ to received number $k - j + 1$. Then Alice generates random prime numbers p of $N/2$ bits and computes the values

$$z_u = y_u (\mathrm{mod}\, p) \text{ for u} \in \{0 \ldots 9\}$$

until all z_u differ by at least 2 modulo p. Then, Alice sends the chosen prime p and the following set of numbers to Bob:

$$z_0, z_1, \ldots z_i, z_i + 1, z_{i+1} + 1, \ldots z_{10} + 1.$$

In other words, Alice increases z_u for all u larger than her value i. From these numbers, Bob can decide whether $i < j$ by looking at the jth number. If it is equal to $z_j = x (\mathrm{mod}\, \mathrm{p})$, then $i \geq j$. If not, then $i < j$.

Algorithm 6 Algorithm for the millionaires problem

Require: $1 < i, j < 10$
 1: Bob: Choose an N-bit random number x.
 2: Bob: Compute $k = \mathcal{E}_a(x)$ and transmit $k - j + 1$
 3: Alice: Calculate $y_u = \mathcal{D}_a(k - j + i)$ for $u \in \{0 \ldots 9\}$
 4: **repeat**
 5:     Alice: Calculate $z_u = y_u (\mathrm{mod}\, p)$ for $u \in \{0 \ldots 9\}$
 6: **until** all z_u differ by at least 2 modulo p.
 7: Alice: Send $p, z_0, z_1, \ldots z_i, z_i + 1, z_{i+1} + 1, \ldots z_{10} + 1$ to Bob
 8: **if** Bob: $z_j == x (\mathrm{mod}\, p)$ **then**
 9:     Bob: Conclude $i \geq j$.
10: **else**
11:     Bob: Conclude $i < j$.
12: **end if**

This algorithm is a beautiful example of how one can hide information due to cryptography and only lack partial information: in essence, Alice destroys the decryptability of all z_i higher than her own value i. Consequently, if $i < j$, Alice will have destroyed the decryptability of z_j and the equality test will fail. If, however, $i \geq j$, then the jth value z_j will not have been hit by Alice and the additions cancel out and the number is correctly decryptable. Hence, the test $z_j == x (\mathrm{mod}\, p)$ of Bob will return the correct answer to the problem.

This problem has motivated a vast amount of research on secure multipart computation. For each specific data type and query type, the question whether a secure multipart computation algorithm exists has been discussed. For location-based services, secure multipart computations are, however, often not feasible as they have some assumptions which are difficult to realize. The given algorithm, for example, is based on enumerating the bounded domain and decrypting for every element of the bounded domain. This becomes quickly infeasible as this computation cannot be cached in any way, because the random seed of Bob trips into the computation. Especially the asymmetric nature of indoor location-based services, namely, that many users use a single service associated with a building, makes this type of algorithm difficult.

Therefore, a lot of database research has been done into providing query-level privacy: assume that the mobile user provides his data to some database, usually trusted third party. How can the database decide which queries compromise a mobile user's privacy? This question has led to a simple and reasonable approach called *k*-anonymity.

9.2 *k*-Anonymity

Privacy in databases is often based on anonymity of records: let A be a database consisting of instances $i_1 \ldots i_N$ associated with attributes $a_1 \ldots a_M$. Each instance, hence, consists of a tuple of M values for the different attributes. Such a database is often represented by a table. Adopting a classical example for *k*-anonymity, we list a small database in Table 9.1.

Table 9.1 Inpatient microdata [6]

	Nonsensitive			Sensitive
	Zip code	Age	Nationality	Condition
1	13053	28	Russian	Heart disease
2	13068	29	American	Heart disease
3	13068	21	Japanese	Viral infection
4	13053	23	American	Viral infection
5	14853	50	Indian	Cancer
6	14853	55	Russian	Heart disease
7	14850	47	American	Viral infection
8	14850	49	American	Viral infection
9	13053	31	American	Cancer
10	13053	37	Indian	Cancer
11	13068	36	Japanese	Cancer
12	13068	35	American	Cancer

This database consists of 12 instances containing three nonsensitive attributes: zip code, age, and nationality. The last column contains the sensitive attribute condition. For the construction of *k*-anonymity, a central assumption is that the instances of the table represent individuals from some population Ω. This is due to the fact that the anonymity in the framework of *k*-anonymity is measured with respect to this population. Some instance of a database is anonymous if it cannot be linked with a single element of the population Ω. For the construction of *k*-anonymity, several specific subsets of the nonsensitive attributes are constructed, which are able to identify an entry of a database inside the total population Ω.

Definition 9.1 A *quasi-identifier* is a set of nonsensitive attributes of a table which can be linked with external data to uniquely identify at least one individual of the population Ω.

A very simple quasi-identifier is a primary key in Ω. A primary key in Ω is an attribute whose value changes for all instances of the population. This could, for example, be the social security number for the population of a single country. In more complicated settings, this can also be the collection of several attributes which together identify a single instance of some population. Let the set of all quasi-identifiers be denoted by QI, that is, the set of all subsets of attributes that can identify at least one instance in the population. With this definition in place, k-anonymity is defined as follows:

Definition 9.2 A table T satisfies k-anonymity if for every instance t there exist at least $k-1$ other instances $t_{a_1}, \ldots, t_{a_{k-1}}$ such that they are indistinguishable when projected onto all subsets of quasi-identifiers $\mathscr{C} \in QI$.

In general, a table such as Table 9.1 does not possess k-anonymity. Hence, the database is not published. Instead, some anonymization procedure is used to generate another table T^* for which k-anonymity holds and some maximum of information of T is kept intact. One such approach is given by generalization: if we split the last digit of age, the number of different age entries decreases from 12 to 4. The idea of k-anonymity is now given by generalizing among the nonsensitive attributes such that in every group of equal nonsensitive attributes, more than k elements are given. As an example, Table 9.2 provides a k-anonymous generalization of Table 9.1 for $k=4$.

Table 9.2 4-Anonymous generalization of inpatient microdata [6]

	Nonsensitive			Sensitive
	Zip code	Age	Nationality	Condition
1	130**	<30	*	Heart disease
2	130**	<30	*	Heart disease
3	130**	<30	*	Viral infection
4	130**	<30	*	Viral infection
5	1485*	≥40	*	Cancer
6	1485*	≥40	*	Heart disease
7	1485*	≥40	*	Viral infection
8	1485*	≥40	*	Viral infection
9	130**	3*	*	Cancer
10	130**	3*	*	Cancer
11	130**	3*	*	Cancer
12	130**	3*	*	Cancer

The algorithmic creation of k-anonymous generalization of datasets is a complex topic in itself and beyond the scope of this chapter. In the 4-anonymous table, for each quasi-identifier subset consisting of zip code, age, and nationality, four different instances of the original database are available. That is, from the generalized table, each instance does not differ in the nonsensitive attributes from three other instances. In other words, querying this database with a given set of nonsensitive attributes does reveal sets of at least four different instances.

Due to the conceptual simplicity and the availability of efficient algorithms for generating k-anonymous datasets out of a given dataset, this privacy concept has quickly gained popularity. Unfortunately, k-anonymity does not provide privacy in all cases. That is, not every k-anonymous generalization of a table should be published.

We consider four weaknesses of k-anonymity in the form of examples:

Example 9.1 (Homogeneity Attack) Alice might know that Bob's record is inside Table 9.2. Assume further that Alice knows the zip code, age, and nationality of Bob. Then Alice knows in which block Bob's instance is inside. Assume it is the last block with instances 9–12. As the sensitive attribute in this group is constant, Alice infers that Bob must suffer from cancer.

At first sight, one could think that this attack is very theoretical in practice. However, if there are only few possible values for the sensitive attributes, the probability that such cells exist is quite high. Moreover, increasing k will increase privacy as the number of elements in each cell increases. At the same time, increasing k will decrease the overall amount of published information and, hence, reduce usefulness. A trivial example would be a generalization which suppresses all nonsensitive attributes and, hence, consists of listing all sensitive attributes. This is quite good from a privacy perspective; however, it does not provide any insight and cannot be queried with nonsensitive attributes. Hence, it does not contain any link between nonsensitive and sensitive attributes.

This motivates the aim of finding groupings of instances such that all sensitive attributes are sufficiently diverse leading to the notion of l-diversity explained in the next section.

Example 9.2 (Background Knowledge Attack) Assume again that Alice knows the nonsensitive attributes of an instance of interest and can thus find out the block of data in Table 9.2 containing the instance of interest. Assuming that Alice knows that the probability of suffering from heart disease is very low for the instance of interest, then Alice can conclude that the instance of interest will have a virus infection with high probability.

These two attacks also serve as example for two strategies of an adversary: positive disclosure and negative disclosure. In the case of the background knowledge attack, Alice was not able to infer, which of the instances is her instance of interest. However, she could easily rule out a lot of other instances reducing the effective number $k - 1$ of indistinguishably equal instances. This type of privacy risk is called *negative disclosure*. Formally, a negative disclosure is given when the adversary can rule out some values of the sensitive attributes and, hence, reduce the effective anonymity. The first attack serves as an example of a *positive disclosure* in which the adversary is able to directly map the nonsensitive attribute values to values of the sensitive attributes.

Example 9.3 (Unsorted Publication) Assume that several k-anonymous tables containing the same individuals in the same ordering are published. This can easily happen when the table is time dependent and is published more often than the individuals inside the table change. Then it could easily happen that algorithms for constructing k-anonymous variants of the database suppress different information. In this case, the attacker can collect the non-suppressed information of both tables possibly breaking k-anonymity.

A simple countermeasure against unsorted publication is given by randomly reordering the table every time it is published. However, even with randomized publication, the k-anonymity can be broken if different tables containing one equal instance are published:

Example 9.4 (Block Intersection) Assume that Alice has enough knowledge to find the block in which the instance of Bob is located in two independent k-anonymous tables. Then Alice can combine the values of non-suppressed information in both tables and reduce the number of candidate instances in both tables breaking k-anonymity. In total, Alice can conclude that Bob's record is located inside the intersection of both blocks.

The latter two examples are important for applications of k-anonymity for location-based services. They represent a very common type of attack against anonymization: linking locally anonymous data with each other to deanonymize an entity.

9.3 l-Diversity

Motivated from the attacks based on limited variability of sensitive attributes inside the blocks of a k-diverse version of the database, a refined measure called l-diversity has been proposed. L-diversity is constructed similar to k-anonymity with an additional constraint on the distribution of sensitive attributes inside each block compared to the distribution of this sensitive attribute in the complete population where the table has been sampled from. For patient data this means that the blocks are compared to the distribution of diseases in general. The following definition gives the exact formulation:

Definition 9.3 Let T^* be some generalized table associated with the database T such that the instances are grouped into sets q^* with equal nonsensitive attributes. Then a block q^* is *l-diverse* if it contains at least l well-represented values of the sensitive attributes. A table is l-diverse if each block of instances with equal nonsensitive attributes is l-diverse.

This definition leaves out a concrete definition of what a well-represented value is. This can be instantiated in several ways. In general, well-represented value shall render homogeneity attacks less probable. From this construction, you can conclude that the difficulty of negative disclosure is not changed at all, except that a different amount of generalization might be needed for *l*-diversity. This is to be expected as there is no way to rule out negative disclosures without assuming something about the knowledge of an attacker. Instead, *l*-diversity increases the protection with respect to positive disclosures by organizing blocks in a way such that the diversity of the sensitive attributes is high enough. In other words, the homogeneity attack given in the example above where Alice can identify the block in which Bob's records have to be and in which this block contains only a single value for the sensitive attribute shall be made harder.

In order to measure the hardness of finding out the sensitive value if an attacker correctly found the right block of data, several definitions for *well-represented* have been given.

One of the most beautiful and simple definitions is given by the amount of information that is leaked about the probability of an instance in some group having some specific sensitive value. The amount of information is linked to the relative fraction of occurrence via the entropy as explained in Sect. 2.1.2. This can directly be used in the following way.

Definition 9.4 A block q^* containing values $s_1 \dots s_m$ for the sensitive attribute is *entropy-l-diverse*, if the entropy of the sensitive values exceeds $\log l$:

$$-\sum_{i=1}^{m} p_i \log(p_i) \geq \log(l) \qquad (9.1)$$

where p_i denotes the relative fraction of s_i inside the block q^*. Let n_i denote the number of occurrences of s_i inside the block q^*. Then the relative fractions are given as

$$p_i = \frac{n_i}{\sum_k n_k}.$$

A database is called *l*-diverse if every block q^* of instances with equal nonsensitive attributes is *l*-diverse.

A consequence from this definition is that each block contains at least *l* different values for the sensitive attribute. This is due to the fact that the left-hand side of Eq. (9.1) represents the amount of information needed to represent the actual values. The right-hand side can be seen as the amount of information needed to represent *l* different values. Hence, the minimal amount of information to represent the set of values for the sensitive attribute has to be larger than the amount of information needed to represent *l* objects.

It is clear from the definition of entropy that splitting a q^*-block into two different blocks reduces the entropy as $-p \log p$ is actually a concave function. Hence, the entropy of the overall table seen as a single block must be larger than $\log l$. This prevents *l*-diversity in some cases. But this is rather a feature than a limitation: if the entropy of the sensitive values of the complete table is quite low, say 90% of the values have the specific value cancer, then a homogeneity attack becomes easy.

Therefore, a different notion of well representation has been introduced called recursive (c, l)-diversity. The main idea is to take over the perspective of an attacker to the system: let again $s_1 \ldots s_m$ denote the different possible values of the sensitive attribute inside a block q^*. Let further n_i denote the number of occurrences of s_i in q^*. Sort these numbers n_i in descending order, the resulting sorted sequence being denoted by $r_1 \ldots r_m$. Now a recursive definition of (c, l)-diversity is given: for $l = 2$, a block q^* is $(c, 2)$-diverse, if

$$r_1 < c(r_2 + \cdots + r_m).$$

The constant c is given and can be used to scale this definition of privacy. This definition is then recursively extended to all integer values for l by saying that a block q^* is (c, l)-diverse if it is possible to eliminate some value of the sensitive attribute in q^* to gain an $(c, l - 1)$-diverse block. Putting this together leads to the following definition:

Definition 9.5 A block q^* containing values $s_1 \ldots s_m$ for the sensitive attribute fulfills *recursive (c, l)-diversity* if

$$r_1 < c(r_l + rl + 1 + \cdots + r_m), \tag{9.2}$$

where r_i denotes the number of occurrences of values for the sensitive attribute in descending order. A database fulfills *recursive (c, l)-diversity* if every block of instances with equal nonsensitive attributes fulfills *recursive (c, l)-diversity*.

The parameter c reflects the distribution of the sensitive values inside each block. If c is chosen very small, the values in blocks are enforced to be equally distributed. On the contrary, when c gets larger, Eq. (9.2) becomes less restrictive.

However, sometimes attribute values are trivially known to the complete population, and hence, a successful homogeneity attack does not leak any unusual information about an individual. If, for example, the hospital is specialized to the treatment of cancer, the publication of this disease does not introduce new information to an attacker who knows that the instance of interest is inside the table of this clinic. Moreover, a difference between the skewness of an attribute value in the population Ω and inside the table can lead to a leak of information.

Example 9.5 (Skewness Attack) Assume that the value of a sensitive attribute is skewly distributed across Ω. Say 99% have one value and only 1% have the other value. Then an optimal entropy-2-diverse block can contain both attribute values with a relative fraction of 50%. This, however, leaks partial information to the attacker. The unlikely value is much more likely for the instance of interest as to be expected from the distribution in Ω.

In order to account for this problem, another refinement for the definition of well representation in the l-diversity framework is given by *positive disclosure*

(c, l)-*diversity* allowing sensitive values of low information gain for an attacker to be included too often with respect to Eq. (9.2). Therefore, a set of values for the attributes is constructed for which positive disclosure is allowed and integrated into the definition of recursive (c, l)-diversity.

Another severe weakness of both k-anonymity and l-diversity is their limitation with respect to linkability of values. If, for example, the sensitive attributes contain several instantiations of a more general concept, they might allow for conclusions about each individual inside the group while seeming different to the framework. This leads to the similarity attack in the following example.

Example 9.6 (Similarity Attack) Assume that the values of a sensitive attribute inside some block are all different but imply some fact about all individuals. For example, the condition in the previous database inside some block could take the values testicular cancer, sterility, and impotence. An attacker can conclude that all instances from this block cannot have offspring. This is a quite severe break of privacy.

This attack is a central and quite general problem about data publishing: computers do not yet use enough knowledge to calculate the amount of information leaked by sets of words in a general and reliable manner. This problem will be kept unsolved unless computers are able to understand and learn in a way comparable to human beings. Semantic techniques are a first step towards these skills; however, they are not yet reliable enough to guarantee detection of information leakage in similarity attacks.

9.4 Spatial and Temporal Cloaking

The concept of k-anonymity has been easily transferred to location-based services. In spatial and temporal cloaking, a central trusted third party collects precise location and time information about entities. This central entity then provides k-anonymous version of this data to a service provider, which provides service information based on the coarse location information provided by the trusted third party and without information about the identity of each specific user.

In this approach, the platform maintains a quadtree splitting space recursively into cells. Figure 9.1 depicts a tree subdivision in equally sized cells.

In a quadtree, each node represents some area. The root node represents the complete area. If the number of elements inside the root cell becomes too high, then this cell is split into smaller cells adding a new level to the tree. In the example of the figure, the first split is splitting the square into two parts: A and the rest. The left part is then further split into two parts: the lower part B and the upper part. This upper part is then again split into four squares: C, D, E, and F. In a quadtree subdivision, each level splits space into four different regions, often called quadrants. The quadrants can be defined geometrically, as in the figure, by splitting

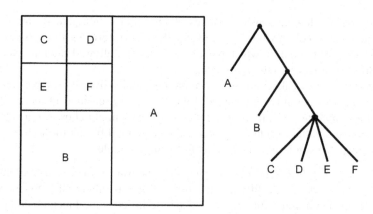

Fig. 9.1 A spatial tree subdivision as used in cloaking approaches

the space into two halves. In other contexts, it is also possible to let the data dictate the splitting by ensuring that the same number of data points is inside each quadrant.

A typical spatial cloaking system starts with having the complete coverage represented by the root node of a quadtree. The mobile devices send location updates and service information at specific times to the central service, which updates the location and membership of each element inside specific quadtree cells. The system ensures that the cells are sufficiently small in order to allow a location-based service to effectively provide service to this area. In a location-based information service, for example, the cells should be small enough such that the transmission of all points of interest inside this cell to the mobile nodes located in these cells becomes feasible.

Location-based services now also connect to this platform and query the platform for specific information about mobile devices such as location, times, or service attributes. The platform now locates the mobile devices inside some leaf node of the tree and uses the tree as a generalization operation: if publishing the requested information with the accuracy given by the current cell does not provide k-anonymity, then the next larger cell is used walking up the tree. If no cell provides anonymity, the service request is rejected. Instead of transmitting the data of the individual for which the service is targeted, the platform will transmit the cell information and a table containing $k - 1$ other instances with the same service information to the service provider. This service provider can then provide the service based on this generalization of location and service information.

Often, the time of occurrence of some service interest or change of location is compromising privacy as well, and this concept of using space division tree in order to have a generalization of location is easily extended to the time domain splitting the time into intervals in which a mobile device has had some specific coarse location. This leads to systems using spatial and temporal cloaking, which provides k-anonymity under the assumptions that the mobile services do not aggregate or reidentify mobile devices, behaviors, or requests.

It has often been shown that approaches such as cloaking are ineffective against the location service provider in practice. Especially when the system is running a longer time, collecting group information and performing intersection attacks are becoming feasible. Though at each instance in time, the system protects privacy, the data from different timestamps can often be correlated. If, for example, the requested service information contains the mobile network operator and k is chosen small, say 2, then repeated requests of a single user lead to anonymization groups whose intersection reveals the network operator of the individual to the service. This is due to the fact that the system cannot ensure that the same instances are always inside the tables reported to location-based services due to scalability problems and the inherent dynamics and mobility of the instances.

9.5 ϵ-Differential Privacy

In the last decade, a new notion of privacy has been established, which is not given as a condition on a table such as k-anonymity or l-diversity. It is rather given as a bound on the variability of the output of a function with respect to small changes on the input to that function. In that, it is quite similar to the notion of a continuous function in math. A continuous function is a function where the amount of change of the function value is bounded by the amount of change of an input to the function. The smaller the input change, the smaller the allowed change of function output. The notion of ϵ-differential privacy is now given with respect to a function evaluation of some function f on several inputs $x_i \in D$ from a common domain D. Given two vectors $x, y \in D^n$, the notion of neighboring vectors is given.

Definition 9.6 Two vectors $x, y \in D^n$ are neighboring if they only differ on exactly one entry.

This definition is independent of the nature of the domain and does not differentiate between small and large changes. However, the amount of change is bounded by the fact that only a single entry of the vectors is allowed to differ.

In order to reflect the distinction between sensitive and nonsensitive attributes in this context, this definition is extended.

Definition 9.7 Let $x, y \in D^n$ be two vectors. Let further $T \subseteq \{1 \dots n\}$ be a set of attribute indices used to select individual entries of the vectors. The two vectors x and y are T-*neighboring* if they only differ on a single entry whose index is not in T.

For the definition of ϵ-differential privacy, the function f of interest must be randomized. This can, for example, be done adding some random noise to f. This randomized function is called $\hat{f}$. The following definition gives the condition on the output of $\hat{f}$ with respect to neighboring vectors (e.g., small changes) of the input.

Definition 9.8 Let $\hat{f} : D^n \to R$ be a randomized function defined on the domain D^n taking values in some set R. Let $\epsilon > 0$. The function $\hat{f}$ is ϵ-differential private if for all neighboring vectors x and y and for all possible sets of outcomes $O \subseteq R$, the following condition is fulfilled:

$$P(\hat{f}(x) \in O) \le e^\epsilon P(\hat{f}(x') \in O),$$

where the probability is calculated with respect to the randomness of $\hat{f}$.

A simple way of ensuring ϵ-differential privacy is given by adding a well-calibrated amount of noise from the Laplace distribution. In this case

$$\hat{f} = f + v,$$

where v is some Laplacian noise calibrated from the global sensitivity of f as follows:

$$v \sim \mathrm{Lap}\left(\frac{\max_{x,y} |f(x) - f(y)|}{\epsilon}\right),$$

where the maximum in the denominator is taken over by all pairs x, y of neighbors in D.

In simple cases, this approach is very easy to apply using a random number generator following a Laplacian distribution. This is best illustrated with a simple example. The following example is taken from [1].

Example 9.7 Let $D = 0,1$ and consider the sum function

$$f(x) = \sum_{i=1}^{n} x_i.$$

Two neighboring vectors in D differ in one entry, hence in the sum, because one entry changes from 0 to 1 or vice versa. Consequently, for neighboring vectors x, y,

$$\max_{x,y} |f(x) - f(y)| = 1.$$

Then $\hat{f}$ is given by adding a noise term v with

$$v \sim \mathrm{Lap}\left(\frac{1}{\epsilon}\right).$$

One of the most important features of ϵ-differential privacy is the careful definition of the term neighboring including a possible set of excluded attributes in the T-neighboring case. A lot of statistical calculations are not sensitive to adding Laplacian noise, as the distribution has zero expectation. Due to this fact, it does, for example, not have a severe impact on the mean or the sum if the number of data points is large enough.

As shown in the example, it is sometimes quite easy to construct ϵ-differential private replacements for functions, especially when the domain of the function is $\mathbb{R}^k$. Because then, the addition of noise is possible. However, it is not clear how to apply this framework, for example, to the medical data from the examples of k-anonymity and l-diversity. How can you add noise to the attribute "Condition?"

In general, ϵ-differential privacy, k-anonymity, and l-diversity are orthogonal approaches. While ϵ-differential privacy is easily applied in continuous domains, k-anonymity and l-diversity expect discrete domains. Still, continuous domains can easily be translated to discrete domains by rounding to a finite number of values making discrete approaches applicable to continuous data.

9.6 Private Information Retrieval

A fundamental different approach to privacy for location-based services is given by private information retrieval. Private information retrieval consists of mechanisms with which a client can query a database without revealing the query contents or the query result to the database. In other words, the query is private. How can such a thing be achieved? From a philosophical point of view, the database has to use all of its data together with the query. If the database is able to leave some data out of query processing, the query will not be completely private, as this negatively discloses some database entries, which are not relevant to the query. This makes private information retrieval complicated and leads to scalability problems. However, private information retrieval can be feasible if the actual privacy demand of the clients is high and the amount of data that needs to be stored in the database is low.

One possibility to implement private information retrieval is based on the assumption that deciding whether a number is a square number some other number is computationally hard. This assumption can be used to formulate a beautiful framework for private information retrieval. The following sections will collect the number of theoretical basement and notations. Then, we will show how the findings of this section can be used to implement private queries.

9.6.1 Quadratic Residues

Let N be some integer number. Then $\mathbb{Z}_N$ denotes the set of residue classes of integer numbers modulo N. The residue classes of this set will be represented by their smallest positive representative, that is, by $0 \ldots N - 1$. We introduce the subset $\mathbb{Z}_N^*$ of numbers coprime with N:

$$\mathbb{Z}_N^* = \{x \in \mathbb{Z}_N \,|\, \gcd(N, x) = 1\}.$$

The numbers in $\mathbb{Z}_N^*$ fall into exactly one of two classes: either they are a quadratic residue modulo N or not. In this context, a quadratic residue is a number y which can be written as a square modulo N. We denote the set of quadratic residues as

$$QR = \{y \in \mathbb{Z}_N^* \,|\, \exists x \in \mathbb{Z}_N^* : y = x^2 \mod N\}.$$

The complement of this set is called the set of quadratic non-residues:

$$QNR = \mathbb{Z}_N^* \setminus QR$$

Table 9.3 Example of quadratic residues modulo 17

a	Relation	Roots	a	Relation	Roots
1	QR	1, 16	10	QNR	–
2	QR	6, 11	11	QNR	–
3	QNR	–	12	QNR	–
4	QR	2, 15	13	QR	8, 9
5	QNR	–	14	QNR	–
6	QNR	–	15	QR	7, 10
7	QNR	–	16	QR	4, 13
8	QR	5, 12	17	QNR	–
9	QR	3, 14			

It is customary to introduce a compact notation for a number being a quadratic residue or not. Therefore, a very classical symbol is given by the Legendre symbol. The Legendre symbol, however, is only defined for residue classes modulo a prime p:

$$\left(\frac{a}{p}\right) = \begin{cases} 1 & \text{if } a \text{ is a quadratic residue modulo } p \\ -1 & \text{if } a \text{ is not a quadratic residue modulo } p \\ 0 & \text{if } p \text{ divides } a \end{cases}.$$

Euler's criterion gives a simple formula for the calculation of the Legendre symbol. Note that it is still limited to p being a prime:

$$\left(\frac{a}{p}\right) = a^{\frac{p-1}{2}} \mod p.$$

The Legendre symbol can be extended to arbitrary numbers N and is renamed Jacobi symbol. However, the notation keeps the same. This symbol is defined recursively given a prime number decomposition of N as follows: let

$$N = p_1^{v_1} p_2^{v_2} \cdots p_k^{v_k}.$$

Then

$$\left(\frac{a}{N}\right) = \left(\frac{a}{p_1}\right)^{v_1} \left(\frac{a}{p_2}\right)^{v_2} \cdots \left(\frac{a}{p_k}\right)^{v_k},$$

where the numbers on the right can be calculated using the Legendre symbol and Euler's criterion. Note that the Jacobi symbol does *not* tell whether a number is a quadratic residue or not.

Example 9.8 As an example to the introduced sets and notations, let us reconsider the notations introduced in some specific cases. Let $p = 17$. Then

$$\mathbb{Z}_N^* = \mathbb{Z}_N \setminus \{0\}.$$

Table 9.3 subsumes the sets QR and QNR in this case. The roots causing the relation are also given.

For private information retrieval, the large number N is now given as the product of two large primes p and q:

$$N = pq.$$

The prime numbers p and q are generated by the client, and the number N is used as a public key. In this situation, the Jacobi symbol can be used to define the subset $\mathbb{Z}_N^+$ of $\mathbb{Z}_N^*$:

$$\mathbb{Z}_N^+ = \left\{ y \in \mathbb{Z}_N^* : \left(\frac{y}{N}\right) = 1 \right\}.$$

This set now consists of quadratic residues modulo N and quadratic non-residues modulo N.

The privacy of private information retrieval is now bound to the hardness of quadratic residuosity assumption, which states that it is computationally hard to find out whether some number is a quadratic residuum modulo N unless the prime decomposition of N is known.

The client will easily find out whether a number is a quadratic residuum modulo $N = pq$ by using Euler's criterion twice:

$$y \in QR \Leftrightarrow \left(y^{\frac{p-1}{2}} == 1 \mod p \vee y^{\frac{q-1}{2}} == 1 \mod q \right).$$

However, for an attacker who does not know the factorization $N = pq$, this is very hard.

9.6.2 Private Information Retrieval Using Quadratic Residuosity

This asymmetric hardness for an attacker and the user can now be exploited to hide a query and an answer from a database. Assume that the database is a bitstring x_i of length L. The server organizes this database as a square matrix $D_{a,b}$ of t rows and t columns. If L is not a proper square, the database will be padded appropriately.

Assume that the client wants to retrieve element x_i, whose index in the square matrix is given by a and b. Then the client generates two large primes p and q for this query and calculates N. Furthermore the client constructs a vector of size t with random entries fulfilling the following condition: every entry of the query vector q_i is a quadratic residue modulo N except the one entry q_b, where the query shall retrieve a bit from the database. The client then sends the number N as well as the query vector q_i to the server.

The server now calculates the following t products to generate a result vector r_i

$$r_i = \prod_{j=1}^{t} \omega_{i,j}.$$

In this equation, $\omega_{i,j} = q_j^2$ if the associated database element $D_{a,b} = 0$ is zero and q_j otherwise. This is the key step of the algorithm: in the row of interest, the server multiplies several elements of QR. As they all have a root modulo N, their product has one, too. Hence, the product keeps inside QR. In one case, by squaring q_j, the server transforms the element from QNR to QR and, hence, multiplies only elements of QR in this row. The answer will contain an element of QR. In the other case, the server introduces q_j, which is in QNR. The product of one element of QNR with elements from QR keeps inside QNR. Hence, the value r_b contains one bit of information, which can be evaluated only by users able to decide whether $r_b \in QR$. This is simple for users knowing the prime factors p and q of N and assumed to be hard otherwise. The server returns this vector in response to the query q_i.

The client now checks the value z_a. Knowing the factorization $N = pq$, he checks whether this is a quadratic residue modulo N or not. If it is, the client deduces that $D_{a,b}$ was zero. Otherwise, $D_{a,b}$ was one.

This completes an algorithm to retrieve one bit of information from a square matrix database based on the property of being a quadratic residuum. Geometrically, the client selects a column of the database $D_{a,b}$ by setting $y_b \in QNR$, and the server returns numbers for each row of the database. The protocol needs $O(n)$ multiplications and $O(\sqrt{n})$ communication.

9.7 Summary

Privacy is very hard to achieve for location-based services and the number and diversity of approaches to this problem are vast. Privacy models such as k-anonymity and l-diversity have problems with external knowledge attackers and are only local in time. One can expect that it is possible to extract non-anonymous location information from several l-diverse tables over time as anonymization is done with respect to the nearest l elements in the database, whose intersection will over time become smaller and smaller. Spatial and temporal cloaking is a specific instantiation of k-anonymity and l-diversity concepts. However, the cloaking regions are hard to protect against semantic knowledge. For example, consider very large cloaking regions of several square kilometers which are empty except for two farms located therein. Is anonymity now felt against the number of people inside this cloaking region, which might be large, or against the number of semantic locations, which is two? Open-ended questions are how to detect such a situation and how to deal with the additional complexity. Another approach to privacy is given by adding noise. Especially the family of ϵ-differential privacy provides a framework for private computation. However, how can we assess the utility of the obscured data to a location-based service? And can we really afford adding additional noise in indoor scenarios? Which and how many location-based services are actually possible for a given ϵ? A map service, for example, will not work. The user wants to see his actual and current location. Private information retrieval can build a bridge towards keeping information local to the device, that is, not disclosing any private information at all. Due to the vast amount of computation and communication needed for real-world applications, this beautiful framework is limited to situations, where the energy demand of such an approach is well invested and the database is small.

Summarizing the findings of this chapter, we can conclude that privacy technologies are available. However, they are not yet ready to be applied in large-scale deployments as they all have severe problems either with unavailable privacy guarantees or with computational and communication overhead. From a user perspective, however, one should also consider the difference of providing a system that claims to provide privacy which does not match our intuitive privacy demands compared to a system that does not protect privacy and communicates this fact clearly and unambiguously. Ineffective privacy protection introduces a feeling of privacy that is not reflected in reality. A very important consideration with respect to privacy could be given by finding cryptographic algorithms for private information retrieval, which are feasible for very large databases and many resource-limited clients.

9.8   Further Reading

Literature about privacy in location-based services is spread out over several different fields of research. For basic privacy with respect to tables including k-anonymity, l-diversity, and t-closeness, still the original references provide the best source of information [5–7]. Tailored to location data, the surveys by Beresford and Stajano [2] and Krumm provide a good overview of the area [4]. For the novel concept of ϵ-differential privacy, I recommend reading the survey of Dwork [3].

References

1. Beimel, A., Nissim, K., Omri, E.: Distributed private data analysis: simultaneously solving how and what. In: Advances in Cryptology—CRYPTO 2008, pp. 451–468. Springer, Berlin (2008)
2. Beresford, A.R., Stajano, F.: Location privacy in pervasive computing. IEEE Pervasive Comput. **2**(1), 46–55 (2003)
3. Dwork, C.: Differential privacy: a survey of results. In: Theory and Applications of Models of Computation, pp. 1–19. Springer, Berlin (2008)
4. Krumm, J.: A survey of computational location privacy. Pers. Ubiquitous Comput. **13**(6), 391–399 (2009)
5. Li, N., Li, T., Venkatasubramanian, S.: t-Closeness: privacy beyond k-anonymity and l-diversity. In: ICDE, vol. 7, pp. 106–115 (2007)
6. Machanavajjhala, A., Kifer, D., Gehrke, J., Venkitasubramaniam, M.: l-Diversity: privacy beyond k-anonymity. ACM Trans. Knowl. Discov. Data **1**(1), 3 (2007)
7. Sweeney, L.: k-Anonymity: a model for protecting privacy. Int. J. Uncertainty Fuzziness Knowl. Based Syst. **10**(05), 557–570 (2002)
8. Yao, A.C.C.: Protocols for secure computations. In: FOCS, vol. 82, pp. 160–164 (1982)

Chapter 10
Open Problem Spaces

*I have yet to see any problem, however complicated, which,
when you looked at it in the right way, did not become still more
complicated.*

Poul Anderson

This chapter gives an outlook towards the next generation of indoor location-based
services. It recollects the organizational structure of this book and explains the
main limitations and open problems for each chapter of this book. This is intended
to motivate researchers to enter some of these topics. This chapter represents the
thoughts of the author and is not representative for the research domain. Still,
the author believes that a lot of people will face the problems listed here when
deploying indoor location-based services and hopes that this orthogonal approach
to the previous chapter is a welcome addition.

10.1 Open Problems in Prerequisites

Chapter 2 entitled Prerequisites contains information about wireless computing
including modulation and signal propagation. While in this basic area, a lot of
developments will be made in the near future, it is unlikely that a fundamental
change will occur with respect to the technical limitations of indoor location-based
services based on wireless signals. Of course, when GNSS signals would become
available, this would add a welcome source of location-dependent information,
and if the bandwidth of communication system increases further, off-loading
and distributed or large-scale computing might become more and more realistic
approaches for deploying advanced algorithms and a platform for cooperative
algorithms.

© Springer International Publishing Switzerland 2014
M. Werner, *Indoor Location-Based Services*, DOI 10.1007/978-3-319-10699-1__10

10.1.1 Sensor and Timing Accuracy

However, the most important open problem lies in the poor quality of sensory of smartphone devices. When the astonishing performance of expensive sensors becomes available to smartphone computing platforms, a boost in location accuracy for indoor location-based services will be achieved. A very special case of this includes exact timings of physical events inside the computer system. It should become understood to provide a timestamp of nanosecond accuracy to any interaction with the physical surroundings including sending a message via an antenna or registering a step or some changing light condition. A lot of techniques are currently limited to specialized hardware with dedicated time measurement units.

10.1.2 Ambient Sensors and Building Automation

Another direction for future developments is given by providing uniform and standardized interfaces for communicating with building infrastructure. In most buildings, a lot of interconnected sensors are already installed in order to control the heating, detect fire, and control lights. These devices can all be extended to provide some useful information for indoor location-based services. However, proprietary communication techniques hinder the global exploitation of already existing infrastructure for indoor location-based services. However, the vision of the Internet of Things is currently gaining more popularity again. It is the vision of all these small devices building one network, the Internet of Things, whose information can be used to provide smart services without the adamant complexities of system integration. This book does not cover the relation between building automation and indoor location-based services due to the fact that there are yet enough examples of successfully integrating building information into indoor location-based services.

10.2 Basic Positioning Techniques

Though positioning techniques based on sensor data are quite elaborated, the integration of different modalities of measurements inside hybrid positioning systems has not yet evolved enough. Furthermore, the combination of different positioning techniques in order to increase positioning quality still leaves a lot of space for improvements. The most important problem with respect to basic positioning techniques, however, is given by an oversimplified quality metric: for an indoor location-based service, the average error as well as the root mean squared error are bad metrics. Positioning research should not focus only on average accuracy but more on accuracy in critical regions, e.g., when changing a direction, leaving a room, or similar. Or the evaluation should be integrated with a risk model for the location-based service. Think, for example, of a guidance application: an expected error of 10 m is not important if the location is inside a large hall of several hundred square meters. However, a location error of 2 m can result into choosing the wrong way near doors or corners. Another direction for improving location accuracy is given by cooperative positioning. The philosophical background of looking for advances by cooperation is given by realizing that a lot of calculations are performed by different nodes in the same way. Furthermore, the sensors of the different nodes measure the same effects from a different perspective including independent noise. If, for example, a particle filter is used to filter the location of several mobile devices and the movement of these mobile devices is constrained by a building, it is very likely that the particles calculated by one of the mobile devices contain valuable information for other mobile devices. A cooperation of multiple mobile devices in the same situation could then reduce the calculational overhead per node and, hence, lead to better results using the same amount of energy and time. Moreover, with cooperative behavior, measurement outliers could be detected and rejected before they harm the accuracy of some positioning system. Given the successful techniques of fingerprinting, especially Wi-Fi fingerprinting, one challenging task is to overcome the problem of database drift. We need algorithms that are able to reliably recalibrate a database using only incoming location queries and which do not generate additional overhead for the system maintainer. In general, it would be great if the calibration effort can be reduced in any way.

10.3 Building Modeling

With respect to building modeling, the most challenging problem lies in closing the gap between complex models, which in theory would enable a vast amount of different services and reality, in which the floorplans are two-dimensional, inconsistent, and outdated. The data acquisition for environmental models must provide reliable, concise, and flexible representations of the surroundings trading off between accuracy and computational overhead induced by the building model. A three-dimensional model at millimeter accuracy might look beautiful; however, these models easily generate very large datasets, and reasoning about them is not feasible on a smartphone platform. Standardization has been providing a lot of stimulation towards the definition and discussion of environmental models. However, the ideas are mostly all deduced from a building perspective. A standardization of possible indoor location-based services and efforts between the poles best representing a building and best supporting indoor location-based services is still missing.

10.4   Position Refinement

The algorithms explained in Chap. 5 are each optimal for some specific case. Hence, there seems to be no open space for innovation. However, there is a very large gap with respect to complexity between the extended Kalman filter and particle filters. From a service quality point of view, particle filter is an optimal choice when it comes to integrating external knowledge such as floorplans into the filtering algorithm. They can be cleanly put into the weighting step of particle filters altering the weight of different particles in order to reflect their probability with respect to external sources of knowledge such as a floorplan. However, medium-scale particle filters are still infeasible for real-time background tracking on a smartphone platform due to their large energy footprint. Despite that, it is possible to deploy small-scale particle filters in smartphones. However, they are often beaten in accuracy by Kalman filter banks or similar advanced filtering systems. An algorithm in between would be very useful: computationally simple and yet able to integrate map matching and other external knowledge as easy as it is possible with particle filters. In general, the integration of real-time information streams containing, for example, context information into the filter process could be a source of innovation in the position refinement area. And similar to positioning, refinement algorithms should be compared with respect to their behavior when they are most needed and not with respect to their average performance.

10.5   Trajectory Computing

Trajectory computing represents a very important area of research. Neither the geometry of trajectories nor the similarity of trajectories has been fully understood today. There are plenty of similarity scores and distance metrics, which enable the comparison of points, trajectories, and other geometric objects. Still, there is a lot of room left for improvements as these distances all cover very different aspects of trajectories. An integrated view of trajectory similarity measures could provide a handy toolkit for developers of indoor location-based services especially because trajectory computing can help in overcoming phases of high uncertainty. Another aspect is the lack of efficient index structures which can be used to query trajectory data. Though there are a lot of extensions to classical index structures such as the 3D-R-tree, all these structures have their specific limitations. The most important limitation is that it seems to be either possible to have an efficient spatial index structure or to keep neighboring samples near to each other inside this spatial index structure. This area of tension provides room for improvements.

10.6 Event Detection

Indoor location-based services have been discussed from a signal processing perspective in the past: as the most important area of improvement is given by location accuracy and tracking performance, this is to be expected. However, indoor location-based services rely much more on accurate location information. An event-driven architecture provides sufficient flexibility and opens up the problems of indoor location-based services to other research domains, especially to the domains related to semantic computing. While we should still work a lot on improving the elements such as positioning, building representation, crowdsourcing, indexing, and reasoning, we should provide our findings and problems in a more general language and extendible frameworks. In order to do that, indoor location-based research should collect possible events and possible problems formulated on these event streams in order to integrate modern reasoning techniques into indoor location-based services.

10.7 Simultaneous Localization and Mapping in Buildings

Simultaneous localization and mapping has been researched for a long time. It is rooted in robotics, where a limited amount of possible movement is combined with a maximal amount of sensory to interact with the surroundings. For indoor location-based services, an important problem is given by collecting enough environmental data inside buildings. The author believes that the key to this problem is given by crowdsourcing. It is very costly and therefore often unrealistic to create highly detailed map information in large scale. Consequently, voluntarism is needed and crowdsourcing could be a tool for that. This implies some challenges especially for simultaneous localization and mapping approaches: can we do SLAM with multiple users all only providing a small amount of information? Can we join all noisy signals measurable by mobile devices into a dataset from which we can gain map information and location? One of the most important open challenges to SLAM in buildings is given by finding suitable methods of measuring movement as well as detecting loops without investing too much energy. Only if we manage to generate the data in the background without affecting the energy level of smartphones too much and without requiring the user to interact with the system in any way that we can make SLAM-based map information ubiquitously available.

10.8 Privacy and Security Considerations

Privacy is an important consideration with respect to indoor location-based services. Possibly, privacy is even more at risk inside buildings, as there is no anonymous location determination technology available in contrast to GPS for the outside space. On the contrary, however, interaction is quite local: if the building itself provides interfaces in order to provide location to the mobile device, this is naturally distributed and it is not unlikely that it will not be communicated to a central database. Moreover, access control to the system can be bound to physical proximity rendering user accounts needless. Therefore, linking interactions of the same user with different buildings gets more difficult. On the contrary, this is not impossible and as long as targeted advertising is the main source of money for large Internet companies, this data is too valuable to be left in distribution. Therefore, one of the most important open challenges is to find a model for a real-world trusted third party. In many cases, the mobile operator would be the perfect entity for providing anonymization as a service: firstly, the operator already knows very much about the location and habits of individual users, and hence, the amount of additional disclosure of private information is small. Secondly, the distributed infrastructure of base stations could be used in order to provide protection against location forgery, which is currently impossible for terminal-based positioning systems.

Another challenge is about the user himself and about communicating the relevance and details of privacy protection to the user. When a user decides to use some application- or location-based service, he or she is mainly interested in the service, say seeing a map of the surroundings or seeing which friends are nearby. The user interface is very small and the topic of privacy and loss of privacy is very complex. It would be optimal if a user was able to protect his privacy without digging into what privacy means and what his personal risk is. In general, privacy protection leads to reduced service quality and negatively impacts the user experience. It is difficult to explain and motivate users to accept this negative impact in reality.

From a more theoretical point of view, it would be important to find new private query protocols, which enable the large-scale application of private queries for location-based datasets. However, the query complexity will always be bound to the size of the database, and consequently databases will have to be split appropriately to find a computationally feasible approach which provides sufficient privacy.

10.9   Summary

Indoor location-based services have reached a stage of development in which exciting and useful services become feasible. However, the deployment and adoption of indoor location-based services are still very limited. This might be due to the case that the expectation of possible service providers orient on outdoor location-based services and do not dare yet to deploy services with much lower service quality as compared to similar outdoor services.

For all aspects of indoor location-based services, it is to be expected that innovative approaches will outperform the classical approaches described in this book. Moreover, none of the algorithms and approaches has reached a level in which there exists no demand for improvement. For all areas covered by this book, a lot of application ideas exist for which algorithms, techniques, and accuracy in measurement are still missing. Therefore, indoor location-based services provide a fascination area of research in which the boundary between classical research areas dissolve.

Index

Printed in the United States
By Bookmasters